Scientific Methods in Mobile Robotics

Ulrich Nehmzow

Scientific Methods in Mobile Robotics

Quantitative Analysis of Agent Behaviour

With 116 Figures

 Springer

Ulrich Nehmzow, Dipl Ing, PhD, CEng, MIEE
Department of Computer Science
University of Essex
Colchester CO4 3SQ
United Kingdom

British Library Cataloguing in Publication Data
Nehmzow, Ulrich, 1961-
    Scientific methods in mobile robotics : quantitative
    analysis of agent behaviour. - (Springer series in advanced
    manufacturing)
    1. Mobile robots 2. Robots - Dynamics - Simulation methods
    I. Title
    629.8'932

ISBN-13: 978-1-84996-543-9   e-ISBN 978-1-84628-260-7   Printed on acid-free paper

Printed in Germany

9 8 7 6 5 4 3 2 1

Springer Science+Business Media
springeronline.com

S.D.G.

Dedicated to the RobotMODIC group:

Steve Billings, Theocharis Kyriacou, Roberto Iglesias Rodríguez,
Keith Walker and Hugo Vieira Neto,

and its support team:

Claudia and Henrietta Nehmzow,
Maria Kyriacou, Michele Vieira and Maxine Walker

Foreword

Mobile robots are widely applied in a range of applications from transportation, surveillance through to health care. In all these applications it is clearly important to be able to analyse and control the performance of the mobile robot and it is therefore surprising that formalised methods to achieve this are not readily available. This book introduces methods and procedures from statistics, dynamical systems theory, and system identification that can be applied to address these important problems. The core objective is to try to explain the interaction between the robot, the task and the environment in a transparent manner such that system characteristics can be analysed, controllers can be designed, and behaviours can be replicated in a systematic and structured manner. This aim of constructing a formalised approach for task-achieving mobile robots represents a refreshingly new approach to this complex set of problems.

Dr Nehmzow has done an outstanding job of constructing and describing a unified framework, which clearly sets out the crucial issues for the development of a theory for mobile robots. Thanks to the careful organisation of the topics and a clear exposition, this book provides an excellent introduction to some new directions in this subject area. Dr Nehmzow's book represents a major departure from the traditional treatment of mobile robots, and provides a refreshing new look at some long-standing problems. I am sure that this is just the beginning of an exciting new phase in this subject area. This book provides a very readable account of the concepts involved; it should have a broad appeal, and will I am sure provide a valuable reference for many years to come.

S A Billings
Sheffield, May 2005

Preface

This book is about scientific method in the investigation of behaviour, where "behaviour" stands for the behaviour of any "behaving" agent, be it living being or machine. It therefore also covers the analysis of robot behaviour, but is not restricted to that. The material discussed in this book has been equally successfully presented to biologists and roboticists alike!

"Scientific method" here stands for the principles and procedures for the systematic pursuit of knowledge [Merriam Webster, 2005], and encompasses the following aspects:

- Recognition and formulation of a problem
- Experimental procedure, consisting of experimental design, procedure for observation, collection of data and interpretation
- The formulation and testing of hypotheses

The hypothesis put forward in this book is that behaviour — mainly motion — can be described and analysed quantitatively, and that these quantitative descriptions can be used to support principled investigation, replication and independent verification of experiments.

This book itself is an experiment. Besides analysing the behaviour of agents, it investigates the question of how ready we are, as a community of robotics practitioners, to extend the practices of robotics research to include *exact* descriptions of robot behaviour, to make testable predictions about it, and to include independent replication and verification of experimental results in our repertoire of standard procedures.

I enjoyed developing the material presented in this book very much. It opened up a new way of doing robotics, led to animated, stimulating and fruitful discussion, and new research (the "Robot Java" presented in Section 6.7 is one example of this). Investigating ways of interpreting experimental results *quantitatively* led to completely new experimental methods in our lab. For example, instead of simply developing a self-charging robot, say, we would try to find the baseline, the "standard" with which to compare our results. This meant that publications would no longer only contain the description of a particular result (an existence proof), but also its quantitative comparison with an established baseline, accepted by the community.

The responses so far to these arguments have been truly surprising! There seems to be little middle ground; the topic of employing scientific methods in robotics appears to divide the community into two distinct camps. We had responses across the whole spectrum: on the one hand, one of the most reputable journals in robotics even denied peer review to a paper on task identification and rejected it without review, and in one seminar the audience literally fell asleep! On the other hand, the same talk given two days later resulted in the request to stay an extra night to "discuss the topic further tomorrow" (and this was after two hours of discussion); the universities of Palermo, Santiago de Compostela and the Memorial University Newfoundland requested "Scientific Methods in Robotics" as an extra mural course, changed the timetables for all their robotics students and examined them on the guest lectures!

I am encouraged by these responses, because they show that the topic of scientific methods in mobile robotics is not bland and arbitrary, but either a red herring or an important extension to our discipline. The purpose of this book is to find out which, and to encourage scientific discussion on this topic that is a principled and systematic engagement with the argument presented. If you enjoy a good argument, I hope you will enjoy this one!

Acknowledgements

Science is never done in isolation, but crucially depends on external input. "As iron sharpens iron, so one man sharpens another" (Prov. 27,17), and this book proves this point. I may have written it, but the experiments and results presented here are the result of collaboration with colleagues all over the world. Many of them have become friends through this collaboration, and I am grateful for all the support and feedback I received.

Most of the experiments discussed in this book were conducted at the University of Essex, where our new robotics research laboratory provided excellent facilities to conduct the research presented in this book. I benefited greatly from the discussions with everyone in the Analytical and Cognitive Robotics Group at Essex — Theo Kyriacou, Hugo Vieira Neto, Libor Spacek, John Ford and Dongbing Gu, to name but a few — as well as with my colleague Jeff Reynolds. Much of this book was actually written while visiting Phillip McKerrow's group at the University of Wollongong; I appreciate their support, and the sabbatical granted by Essex University. And talking of sabbaticals, Keith Walker (Point Loma Nazarene University, San Diego) and Roberto Iglesias Rodriguez (Dept. of Electronics and Computer Science at the University of Santiago de Compostela) made important contributions during their sabbaticals at Essex. I am also indebted to many colleagues from other disciplines, notably the life sciences, who commented on the applicability of methods proposed in this book to biology, psychology *etc*. I am especially grateful for the support I received from Wolfgang and Roswitha Wiltschko and their group at the J.W. Goethe University in Frankfurt.

The RobotMODIC project, which forms the backbone of work discussed in this book, would not have happened without the help and commitment of my colleague and friend Steve Billings at the University of Sheffield, the committed work by my colleague and friend Theo Kyriacou, and the support by the British Engineering and Physical Sciences Research Council. I benefited greatly from all this scientific, technical, financial and moral support, and thank my colleagues and sponsors.

Finally, I thank all my family in Germany for their faithful, kind and generous support and love. My wife Claudia, as with book #1, was a constructive help all along the way, and Henrietta was a joy to be "criticised" by. Thank you all!

As before, I have written this book with Johann Sebastian Bach's motto "SDG" firmly in mind.

Ulrich Nehmzow
Colchester, Essex, October 2005

Contents

1 A Brief Introduction to Mobile Robotics 1
  1.1 This Book is not about Mobile Robotics 1
  1.2 What is Mobile Robotics? 1
  1.3 The Emergence of Behaviour 5
  1.4 Examples of Research Issues in Autonomous Mobile Robotics . . 7
  1.5 Summary 9

2 Introduction to Scientific Methods in Mobile Robotics 11
  2.1 Introduction 11
  2.2 Motivation: Analytical Robotics 13
  2.3 Robot-Environment Interaction as Computation 15
  2.4 A Theory of Robot-Environment Interaction 16
  2.5 Robot Engineering *vs* Robot Science 18
  2.6 Scientific Method and Autonomous Mobile Robotics 19
  2.7 Tools Used in this Book 27
  2.8 Summary: The Contrast Between
      Experimental Mobile Robotics and Scientific Mobile Robotics . . 28

3 Statistical Tools for Describing Experimental Data 29
  3.1 Introduction 29
  3.2 The Normal Distribution 30
  3.3 Parametric Methods to Compare Samples 33
  3.4 Non-Parametric Methods to Compare Samples 43
  3.5 Testing for Randomness in a Sequence 55
  3.6 Parametric Tests for a Trend (Correlation Analysis) 57
  3.7 Non-Parametric Tests for a Trend 65
  3.8 Analysing Categorical Data 69
  3.9 Principal Component Analysis 80

4   Dynamical Systems Theory and Agent Behaviour 85
    4.1   Introduction 85
    4.2   Dynamical Systems Theory 85
    4.3   Describing (Robot) Behaviour Quantitatively Through Phase
          Space Analysis 95
    4.4   Sensitivity to Initial Conditions: The Lyapunov Exponent 100
    4.5   Aperiodicity: The Dimension of Attractors 116
    4.6   Summary 119

5   Analysis of Agent Behaviour — Case Studies 121
    5.1   Analysing the Movement of a Random-Walk Mobile Robot 121
    5.2   "Chaos Walker" 126
    5.3   Analysing the Flight Paths of Carrier Pigeons 133

6   Computer Modelling of Robot-Environment Interaction 139
    6.1   Introduction 139
    6.2   Some Practical Considerations Regarding Robot Modelling 141
    6.3   Case Study: Model Acquisition Using Artificial Neural Networks 143
    6.4   Linear Polynomial Models and Linear Recurrence Relations 150
    6.5   NARMAX Modelling 155
    6.6   Accurate Simulation: Environment Identification 156
    6.7   Task Identification 173
    6.8   Sensor Identification 184
    6.9   When Are Two Behaviours the Same? 185
    6.10   Conclusion 193

7   Conclusion 195
    7.1   Motivation 195
    7.2   Quantitative Descriptions of Robot-Environment Interaction 196
    7.3   A Theory of Robot-Environment Interaction 197
    7.4   Outlook: Towards Analytical Robotics 199

References 201

Index 205

1

A Brief Introduction to Mobile Robotics

Summary. This chapter gives a brief introduction to mobile robotics, in order to set the scene for those readers who are not familiar with the area.

1.1 This Book is not about Mobile Robotics

This book is not actually about mobile robotics! It is merely written from a mobile robotics perspective, and the examples given are drawn from mobile robotics, but the question it addresses is that of "analysing behaviour", where behaviour is a very loose concept that could refer to the motion of a mobile robot, the trajectory of a robot arm, a rat negotiating a maze, a carrier pigeon flying home, traffic on a motorway or traffic on a data network. In short, this book is concerned with describing the behaviour of a dynamical system, be it physical or simulated. Its goals are to analyse that behaviour quantitatively, to compare behaviours, construct models and to make predictions. The material presented in this book should therefore be relevant not only to roboticists, but also to psychologists, biologists, engineers, physicists and computer scientists.

Nevertheless, because the examples given in this book are taken from the area of mobile robotics, it is sensible to give at least a very brief introduction to mobile robotics for the benefit of all the non-roboticists reading this book. A full discussion of mobile robotics is found in [Nehmzow, 2003a], and if this book is used as teaching material, it is advisable students read general introductions to mobile robotics such as [Nehmzow, 2003a, Siegwart and Nourbakhsh, 2004, Murphy, 2000] first.

1.2 What is Mobile Robotics?

Figure 1.1 shows a typical mobile robot, the Magellan Pro *Radix* that is used at the University of Essex. Equipped with over 50 on-board sensors and an on-board computer, the robot is able to perceive its environment through its sensors, process the signals on its computer, and as a result of that computation control its own motion through space.

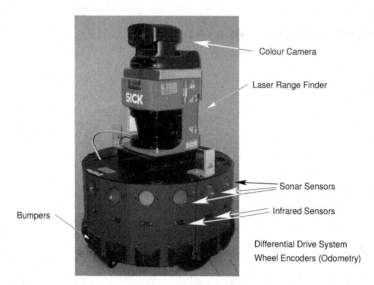

Figure 1.1. *Radix*, the Magellan Pro mobile robot used in the experiments discussed in this book

Radix is completely autonomous, meaning that it is not dependent upon any link to the outside world: it carries its own batteries, therefore not needing an umbilical cord to supply power, and has its own computer, therefore not needing a cable or radio link to an external control mechanism. It is also not remote-controlled by a human, but interacts with its environment autonomously, and determines its motion without external intervention.

Not all mobile robots are autonomous, but all mobile robots are capable of moving between locations. This might be achieved using legs or wheels and there are mobile robots that can climb walls, swim or fly. The discipline of mobile robotics is concerned with the control of such robots: how can the task they are designed for be achieved? How can they be made to operate reliably, under a wide range of environmental conditions, in the presence of sensor noise, contradictory or erroneous sensor information? These are the kinds of questions mobile robotics addresses.

1.2.1 Engineering

Obviously, a mobile robot is made up of hardware: sensors, actuators, power supplies, computing hardware, signal processing hardware, communication hardware, *etc*. This means that there is a strong engineering element in designing mobile robots, and a vast amount of background literature exists about the engineering aspects of robotics [Critchlow, 1985, McKerrow, 1991, Fuller, 1999, Martin, 2001]. Journals addressing the engineering aspects of robotics include, among many more, *Advanced Robotics, Automation in Construction, Industrial Robot, IEEE Trans. on Robotics, IEEE Trans. on Automation Science and Engineering, International Journal of Robotics Research, Journal of Intelligent and Robotic Systems, Mechatronics, Robotica, Robotics and Autonomous Systems* and *Robotics and Computer Integrated Manufacturing*.

1.2.2 Science

An autonomous mobile robot closes the loop between perception and action: it is capable of perceiving its environment through its sensors, processing that information using its on-board computer, and responding to it through movement. This raises some interesting questions, for example the question of how to achieve "intelligent" behaviour. What are the foundations of task-achieving behaviours, by what mechanism can behaviours be achieved that appear "intelligent" to the human observer? Second, there is a clear parallel between a robot's interaction with the environment and that of animals. Can we copy animal behaviour to make robots more successful? Can we throw light on the mechanisms governing animal behaviour, using robots?

Such questions concerning behaviour, traditionally the domain of psychologists, ethologists and biologists, we refer to as "science". They are not questions of hardware and software design, *i.e.* questions that concern the robot itself, but questions that *use* the mobile as a tool to investigate other questions. Such use of mobile robots is continuously increasing, and a wide body of literature exists in this area, ranging from "abstract" discussions of autonomous agents ([Braitenberg, 1987, Steels, 1995, von Randow, 1997, Ritter et al., 2000]) to the application of Artificial Intelligence and Cognitive Science to robotics ([Kurz, 1994, Arkin, 1998, Murphy, 2000]
[Dudek and Jenkin, 2000]). Journals such as *Adaptive Behavior* or *IEEE Transactions on Systems, Man, and Cybernetics* also address issues relevant to this topic.

1.2.3 (Commercial) Applications

Mobile robots have fundamental strengths, which make them an attractive option for many commercial applications, including transportation, inspection, surveillance, health care [Katevas, 2001], remote handling, and specialist applications like operation in hazardous environments, entertainment robots ("artificial pets") or even museum tour guides [Burgard et al., 1998].

Like any robot, mobile or fixed, mobile robots can operate under hostile conditions, continuously, without fatigue. This allows operation under radiation, extreme temperatures, toxic gases, extreme pressures or other hazards. Because of their capability to operate without interruption, 24 h of every day of the week, even very high investments can be recovered relatively quickly, and a robot's ability to operate without fatigue reduces the risk of errors.

In addition to these strengths, which all robots share, mobile robots have the additional advantage of being able to position themselves. They can therefore attain an optimal working location for the task at hand, and change that position during operation if required (this is relevant, for instance, for the assembly of large structures). Because they can carry a payload, they are extremely flexible: mobile robots, combined with an on-board manipulator arm can carry a range of tools and change them on site, depending on job requirements. They can carry measurement instruments and apply them at specific locations as required (for example measuring temperature, pressure, humidity *etc.* at a precisely defined location). This is exploited, for instance, in space exploration [Iagnemma and Dubowsky, 2004].

Furthermore, cooperative mobile robot systems can achieve tasks that are not attainable by one machine alone, for example tasks that require holding an item in place for welding, laying cables or pipework, *etc.* Cooperative robotics is therefore a thriving field of research. [Beni and Wang, 1989, Ueyama et al., 1992] [Kube and Zhang, 1992,   Arkin and Hobbs, 1992,   Mataric, 1994]   and [Parker, 1994] are examples of research in this area.

There are also some weaknesses unique to mobile robots, which may affect their use in industrial application.

First, a mobile robot's distinct advantage of being able to position itself introduces the weakness of reduced precision. Although both manipulators and mobile robots are subject to sensor and actuator noise, a mobile robot's position is not as precisely defined as it is in a manipulator that is fixed to a permanent location, due to the additional imprecision introduced by the robot's chassis movement. Furthermore, any drive system has a certain amount of play, which affects the theoretical limits of precision.

Second, there is an element of unpredictability in mobile robots, particularly if they are autonomous, by which is meant the ability to operate without external links (such as power or control). With our current knowledge of the process of robot-environment interaction it is not possible to determine stability limits and behaviour under extreme conditions analytically. One of the aims of this book is to develop a theory of robot-environment interaction, which would allow a theoretical analysis of the robot's operation, for example regarding stability and behaviour under extreme conditions.

Third, the payload of any mobile robot is limited, which has consequences for on-board power supplies and operation times. The highest energy density is currently achieved with internal combustion engines, which cannot be used in many application scenarios, for example indoors. The alternative, electric actuation, is dependent on either external power supplies, which counteract the inherent advantages of mobility because they restrict the robot's range, or on-board batteries, which currently are very heavy. As technology progresses, however, this disadvantage will become less and less pronounced.

1.3 The Emergence of Behaviour

Why is it that a mobile robot, programmed in a certain way and placed in some environment to execute that program, behaves in the way it does? Why does it follow exactly the trajectory it is following, and not another?

The behaviour of a mobile robot — what is observed when the robot interacts with its environment — is not the result of the robot's programming alone, but results from the makeup of three fundamental components:

1. The program running on the robot (the "task")
2. The physical makeup of the robot (the way its sensors and motors work, battery charge, *etc*)
3. The environment itself (how visible objects are to the robot's sensors, how good the wheel grip is, *etc*)

The robot's behaviour *emerges* from the interaction between these three fundamental components. This is illustrated in Figure 1.2.

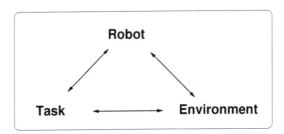

Figure 1.2. The fundamental triangle of robot-environment interaction

This point is easily illustrated. That the robot's behaviour changes when its control program changes is obvious. But likewise, take an "obstacle avoiding" mobile robot, and dump it in a swimming pool! Clearly, what was meant by "obstacle avoiding" was "obstacle avoiding in such and such an environment". Finally, change the robot's sensors, for example by unplugging one sensor, and the behaviour will change as well. When talking about robot behaviour, it is essential to talk about task, robot and environment at the same time. The purpose of scientific methods in mobile robotics is to analyse and understand the relationship between these three fundamental components of the generation of behaviour.

1.3.1 What Makes Robotics Hard?

A mobile robot is an embedded, situated agent. Embedded, because it interacts with its environment through its actions, situated, because its actions affect future states it will be in. And unlike computer simulations (even those involving pseudo random numbers) the interaction between a robot and its surroundings is not always predictable, due to sensor and actuator noise, and chaos inherent in many dynamical systems. What differentiates a physical mobile robot, operating in the real world from, for example, a computer simulation, is the issue of repeatability: if desired, the computer simulation can be repeated exactly, again and again. In a mobile robot, this is impossible.

Figure 1.3 shows the results of a very simple experiment that was designed to illustrate this phenomenon. A mobile robot was placed twice at the same starting location (as much as this was possible), executing the same program in the same environment. Both runs of what constitutes the same experiment were run within minutes of each other.

As can be seen from Figure 1.3, the two trajectories start out very similar to each other, but after two or three turns diverge from each other noticeably. Very shortly into the experiment the two trajectories are very different, although nothing was changed in the experimental setup! The robot is unchanged, the task is unchanged, and the environment is unchanged. The only difference is the starting position of the robot, which differs very slightly between the two runs.

The explanation of this surprising divergence of the two trajectories is that small perturbations (*e.g.* sensor noise) quickly add up, because a slightly different perception will lead to a slightly different motor response, which in turn leads to another different perception, and so on, so that soon two different trajectories emerge. It is this behaviour (which can be "chaotic", see Chapter 4) that makes "real world" robotics so difficult to model, and which leads to pronounced differences between the predictions of a computer simulation and the behaviour of the actual robot. This is not a fault of the robot, but "a natural and proper part of the robot-environment interaction. ... Behaviour is not a property of an agent, it is a dynamical process constituted of the interactions between an agent and its environment" [Smithers, 1995].

Figure 1.4 shows the phenomenon observed during a "real world" experiment, which was actually concerned with the robot exploring the environment over a period of time. During the robot's exploration, it happened to visit the location indicated with "Start" twice, at different moments in time. Initially, the two trajectories follow each other closely, but the first, small divergence is observed at the first turn (point "A"). At the second turn ("B"), the divergence is amplified, and at point "C" the initially close trajectories have diverged so far from each other that the robot takes radically different actions in each case! The trajectory shown as a solid line turns out not to be repeatable.

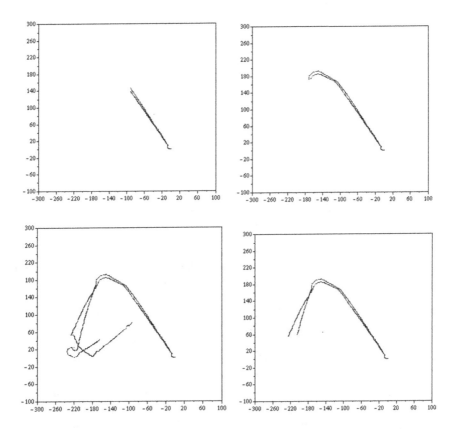

Figure 1.3. The behaviour of a mobile robot is not always predictable. Figures show trajectories over time, clockwise from the *top left* diagram

1.4 Examples of Research Issues in Autonomous Mobile Robotics

The purpose of the concluding section of this chapter is to highlight a few areas where mobile robots are used, by way of example. This section is not comprehensive, but merely aims to give a "feel" of what is being done in mobile robotics. For a more detailed presentation of topics, see textbooks like [Arkin, 1998, Murphy, 2000] and [Nehmzow, 2003a].

1.4.1 Navigation

The advantages of mobility cannot be fully exploited without the capability of navigating, and for example in the realm of living beings one would be hard pressed to find an animal that can move but doesn't have some kind of navigational skill. As a consequence, navigation is an important topic in mobile robotics, and attracts much attention.

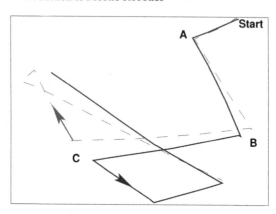

Figure 1.4. Two trajectories observed in a "real world" experiment that set out close to each other, but diverge within a few tens of seconds

Map-based navigation can be defined as the presence of all or at least some of the following capabilities [Nehmzow, 2003a, Nehmzow, 2003b]:

- Self-localisation: without being able to identify one's own position on a map, any navigation is impossible. Self-localisation is the foundation of all navigation.
- Map building: the term "map" here stands for a bijection between two spaces A and B, with A and B not being restricted to navigational maps, but any one-to-one mapping between two spaces (*e.g.* sensory perception and the response of an artificial neural network).
- Map interpretation: the map is of no use to the agent if it is uninterpretable, and map interpretation therefore goes hand in hand with the ability to acquire maps.
- Path planning: this refers to the ability to decide on a sequence of actions that will take the robot from one location to another, and usually involves at least self-localisation and map interpretation.
- Recovery: as stated above, interaction with the real world is partially unpredictable, and any navigating robot needs the ability to recover from error. This usually involves renewed self-localisation and path planning, but sometimes also special recovery strategies, like returning to a known, fixed spot, and navigating anew from there.

Navigational methods applied in mobile robotics broadly encompass mechanisms that use global (often metric) reference frames, using odometry and metric maps.

1.4.2 Learning

In a mobile robot the loop of perception, reasoning and response is closed; mobile robots therefore are ideal tools to investigate "intelligent behaviour". One phenomenon that is frequently observed in nature, and increasingly modelled using mobile robots, is that of learning, *i.e.* the adaptation of behaviour in the light of experience.

The literature in the field of robot learning is vast, for introductions see for instance [Franklin, 1996, Dorigo and Colombetti, 1997, Morik, 1999] [Demiris and Birk, 2000] and [Wyatt and Demiris, 2000].

1.5 Summary

Mobile robotics is a discipline that is concerned with designing the hardware and software of mobile robots such that the robots are able to perform their task in the presence of noise, contradictory and inconsistent sensor information, and possibly in dynamic environments. Mobile robots may be remote controlled, guided by specially designed environments (beacons, bar codes, induction loops *etc.*) or fully autonomous, *i.e.* independent from any links to the outside world.

Mobile robots are widely used in industrial applications, including transportation, inspection, exploration or manipulation tasks. What makes them interesting to scientific applications is the fact that they close the loop between perception and action, and can therefore be used as tools to investigate task-achieving (intelligent) behaviour.

The behaviour of a mobile robot — what is observed when the robot operates — emerges from the interaction between robot, task and environment: the robot's behaviour will change if the robot's hardware is changed, or if the control program (the task) is changed, or if the environment is changed. For example, an unsuccessful wall following robot can be changed into a successful one by either changing the robot's sensors, by improving the control code, or by placing reflective strips on the walls!

The fundamental principles that govern this interaction between robot, task and environment are, at the moment, only partially understood. For this reason it is currently not possible to design mobile robot controllers off line, *i.e.* without testing the real robot in the target environment, and fine tuning the interaction through trial and error. One aim in mobile robotics research, and of this book, therefore is to analyse the interaction between robot, task and environment quantitatively, to gain a theoretical understanding of this interaction which would ultimately allow off-line design of robot controllers, as well as a *quantitative* description of experiments and their results.

2

Introduction to Scientific Methods in Mobile Robotics

Summary. This chapter introduces the main topic of this book, identifies the aims and objectives and describes the background the material presented in this book.

2.1 Introduction

The behaviour of a mobile robot emerges from the relationship and interaction between the robot's control code, the environment the robot is operating in, and the physical makeup of the robot. Change any of these components, and the behaviour of the robot will change.

This book is concerned with how to characterise and model, "identify", the behaviour emerging from the interaction of these three components. Is the robot's behaviour predictable, can it be modelled, is it stable? Is this behaviour different from that one, or is there no significant difference between them? Which programs performs better (where "better" is some measurable criterion)?

To answer these questions, we use methods taken from dynamical systems theory, statistics, and system identification. These methods investigate the dynamics of robot-environment interaction, and while this interaction *is* also governed by the control program being executed by the robot, they are not suited to analyse *all* aspects of robot behaviour. For example, dynamical systems theory will probably not characterise the relevant aspects of the behaviour of a robot that uses computer vision and internal models to steer towards one particular location in the world. In other words, the methods presented in this book are primarily concerned with dynamics, not with cognitive aspects of robot behaviour.

This book aims to extend the way we conduct autonomous mobile robotics research, to add a further dimension: from a discipline that largely uses iterative refinement and trial-and-error methods to one that is based on testable hypotheses, that makes predictions about robot behaviour based on a theory of robot-environment interaction. The book investigates the mechanisms that give rise to robot behaviour we observe: why does a robot succeed in certain environments and fail in others? Can we make accurate predictions as to what the robot is going to do? Can we *measure* robot behaviour?

Although primarily concerned with physical mobile robots, operating in the real world, the mechanisms discussed in this book can be applied to all kinds of "behaving agents", be it software agents, or animals. The underlying questions in all cases are the same: can the behaviour of the agent be *measured quantitatively*, can it be modelled, and can it be predicted?

2.1.1 A Lecture Plan

This book is the result of undergraduate and postgraduate courses in "Scientific Methods in Mobile Robotics" taught at the University of Essex, the Memorial University of Newfoundland, the University of Palermo and the University of Santiago de Compostela. The objective of these courses was to introduce students to fundamental concepts in scientific research, to build up knowledge of the relevant concepts in philosophy of science, experimental design and procedure, robotics and scientific analysis, and to apply these specifically to the area of autonomous mobile robotics research. Perhaps it is easiest to highlight the topics covered in this book through this sequence of lectures, which has worked well in practice:

1. Introduction (Chapter 2):
   - Why is scientific method relevant to robotics? How can it be applied to autonomous mobile robotics?
   - The robot as an analog computer (Section 2.3)
   - A theory of robot-environment interaction (Section 2.4)
   - The role of quantitative descriptions (Section 2.4.2)
   - Robot engineering vs robot science (Section 2.5)
2. Scientific Method (Section 2.6):
   - Forming hypotheses (Section 2.6.2)
   - Experimental design (Section 2.6.3)
   - Traps, pitfalls and countermeasures (Section 2.6.3)
3. Introduction to statistical descriptions of robot-environment interaction:
   - Normal distribution (Sections 3.2 and 3.3.2)
4. Parametric tests to compare distributions:
   - T-test (Sections 3.3.4 and 3.3.5)
   - ANOVA (Section 3.3.6)
5. Non-parametric tests I:
   - Median and confidence interval (Section 3.4.1)
   - Mann-Whitney U-test (Section 3.4.2)
6. Non-parametric tests II:
   - Wilcoxon test for paired observations (Section 3.4.3)
   - Kruskal-Wallis test (Section 3.4.4)
   - Testing for randomness (Section 3.5)

7. Tests for a trend:
   - Linear regression (Section 3.6.1)
   - Pearson's r (Section 3.6.2)
   - Spearman rank correlation (Section 3.7.1)
8. Analysing categorical data (Section 3.8):
   - χ^2 analysis (Section 3.8.1)
   - Cramer's V (Section 3.8.2)
   - Entropy based methods (Section 3.8.3)
9. Dynamical systems theory and chaos theory (Chapter 4):
   - Phase space (Section 4.2.1)
   - Degrees of freedom of a mobile robot (Section 4.2.1)
   - The use of quantitative descriptions of phase space in robotics (Section 2.4.2)
   - Reconstruction of phase space through time-lag embedding (Section 4.2.3)
10. Describing robot behaviour quantitatively through phase space analysis (Section 4.3)
11. Quantitative descriptors of attractors:
    - Lyapunov exponent (Section 4.4)
    - Prediction horizon (Section 4.4.2)
    - Correlation dimension (Section 4.5)
12. Modelling of robot-environment interaction (Chapter 6)
13. ARMAX modelling (Section 6.4.3)
14. NARMAX modelling (Section 6.5):
    - Environment identification (Section 6.6)
    - Task identification (Section 6.7)
    - Sensor identification (Section 6.8)
15. Comparison of behaviours (Section 6.9)
16. Summary and conclusion (Chapter 7)

2.2 Motivation: Analytical Robotics

The aim of this book is to throw some light light on the question "what happens when a mobile robot — or in fact any agent — interacts with its environment?". Can predictions be made about this interaction? If models *can* be built, can they be used to design autonomous mobile robots off-line, like we are now able to design buildings, electronic circuits or chemical compounds without applying trial-and-error methods? Can models be built, and can they be used to hypothesise about the nature of the interaction? Is the process of robot-environment interaction stochastic or deterministic?

Why are such questions relevant? Modern mobile robotics, using autonomous mobile robot with their own on-board power supply, sensors and computing equipment, is a relatively new discipline. While as early as 1918 a light-seeking robot was built by John Hays Hammond [Loeb, 1918, chapter 6], and W. Grey Walter built mobile robots that *learnt* to move towards a light source by way of instrumental conditioning in the 1950s [Walter, 1950, Walter, 1951], "mass" mobile robotics really only began in the 1980s. As in all new disciplines, the focus was initially on the engineering aspects of getting a robot to work: which sensors can be used in a particular task, how do they need to be preprocessed and interpreted, which control mechanism should be used, *etc.* The experimental scenario used was often one of iterative refinement: a good first guess at a feasible control strategy was implemented, then tested in the target environment. If the robot got stuck, failed at the task *etc.*, the control code would be refined, then the process would be repeated until the specified task was successfully completed in the target environment.

A solution obtained in this manner constituted an "existence proof" — it was proven that a particular robot could achieve a particular task under a particular set of environmental conditions. These existence proofs were good achievements, because they demonstrated clearly that a particular behaviour or competence could be achieved, but they lacked one important property: generality. That a robot could successfully complete a navigational route in one environment did not imply that it could do it anywhere else. Furthermore, the experimenter did not really know *why* the robot succeeded. Success or failure could not be determined to a high degree of certainty *before* an experiment. Unlike building bridges, for instance, where civil engineers are able to predict the bridge's behaviour before it is even built, roboticists are unable to predict a robot's behaviour before it is tested.

Perhaps the time has come for us to be able to make some more general, theoretical statements about what happens in robot-environment interaction. We have sophisticated tools such as computer models (see Chapter 6) and analysis methods (see Chapter 4), which can be used to develop a *theory* of robot-environment interaction. If this research wasn't so practical, involving physical mobile robots doing something in the real world, I would call the discipline "theoretical robotics". Instead, I use the term "analytical robotics".

In addition there are benefits to be had from a theory of robot-environment interaction: the more theoretical knowledge we have about robot-environment interaction, the more accurate, reliable and cheap will the robot and controller design process be. The more we know about robot-environment interaction, the more focused and precise will our hypotheses and predictions be about the outcome of experiments. This, in turn, will increase our ability to detect rogue experimental results and to improve our experimental design. Finally, the better understood the process of robot-environment interaction, the better we are able to report experimental results, which in turn supports independent replication and verification of results: robotics would advance from an experimental discipline to one that embraces scientific method.

The aim of this book, therefore, is to understand robot-environment interaction more clearly, and to present abstracted, generalised representations of that interaction — a theory of robot-environment interaction.

2.3 Robot-Environment Interaction as Computation

The behaviour of a mobile robot cannot be discussed in isolation: it is the result of properties of the robot itself (physical aspects — the "embodiment"), the environment ("situatedness"), and the control program (the "task") the robot is executing (see Figure 2.1). This triangle of robot, task and environment constitutes a complex, interacting system, whose analysis is the purpose of any theory of robot-environment interaction.

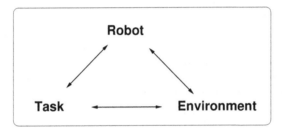

Figure 2.1. The fundamental triangle of robot-environment interaction

Rather than speaking solely of a robot's behaviour, it is therefore necessary to speak of *robot-environment interaction*, and the robot's behaviour resulting thereof.

A mobile robot, interacting with its environment, can be viewed as performing "computation", "computing" *behaviour* (the output) from the three inputs *robot morphology*, *environmental characteristics* and *executed task* (see Figure 2.2).

Similar to a cylindrical lens, which can be used to perform an analog computation, highlighting vertical edges and suppressing horizontal ones, or a camera lens computing a Fourier transform by analog means, a robot's behaviour — for the purposes of this book, and as a first approximation, the mobile robot's trajectory — can be seen as emergent from the three components shown in Figure 2.1: the robot "computes" its behaviour from its own makeup, the world's makeup, and taking into account the program it is currently running (the task).

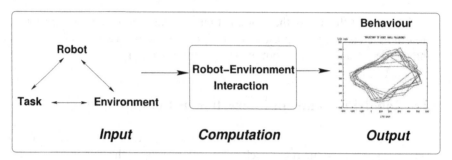

Figure 2.2. Robot-environment interaction as computation: Behaviour (the output) is computed from the three inputs robot morphology, task and environmental properties

2.4 A Theory of Robot-Environment Interaction

2.4.1 Definition

When referring to "theory", we mean a coherent body of hypothetical, conceptual and pragmatic generalisations and principles that form the general frame of reference within which mobile robotics research is conducted.

There are two key elements that make a theory of robot-environment interaction useful, and therefore desirable for research:

1. A theory will allow the formulation of hypotheses for testing. This is an essential component in the conduct of "normal science" [Kuhn, 1964].
2. A theory will make predictions (for instance regarding the outcome of experiments), and thus serve as a safeguard against unfounded or weakly supported assumptions.

A theory retains, in abstraction and generalisation, the essence of what it is that the triple of robot-task-environment does. This generalisation is essential; it highlights the important aspects of robot-environment interaction, while suppressing unimportant ones. Finally, the validity of a theory (or otherwise) can then be established by evaluating the predictions made applying the theory.

Having theoretical understanding of a scientific discipline has many advantages. The main ones are that a theory allows the generation of hypotheses and making testable predictions, but there are practical advantages, too, particularly for a discipline that involves the design of technical artefacts. For instance, theory supports off-line design, *i.e.* the design of technical artefacts through the use of computer models, simulations and theory-based calculations.

2.4.2 The Role of Quantitative Descriptions of Robot-Environment Interaction

Measurement is the backbone of science, and supports:

- The precise documentation of experimental setups and experimental results
- The principled modification of experimental parameters
- Independent verification of experimental results
- Theoretical design of artefacts without experimental development
- Predictions about the behaviour of the system under investigation

We have argued that robot behaviour emerges from the interaction between robot, task and environment. Suppose we were able to measure this behaviour quantitatively. Then, if any two of the three components shown in Figure 2.1 remain unaltered, the quantitative performance measure will characterise the third, modified component. This would allow the investigation of, for instance:

- The effect of modifications of the robot
- The influence of the robot control program on robot behaviour
- The effect of modifications to the environment on the overall behaviour of the robot

This is illustrated in Figure 2.3: the quantitative measure of the robot's behaviour (the dependent variable) changes as some experimental parameter (the independent variable) changes, and can therefore be used to describe the independent variable. For the point γ in Figure 2.3, for example, the quantitative performance measure has a global maximum.

Chapter 4 in particular addresses the question of how robot-environment interaction can be characterised quantitatively, and how such quantitative measures can be used to determine the influence of i) a change in the robot controller, and ii) a change of environment.

Current mobile robotics research practice not only differs from that of established disciplines in its lack of theories supporting design, but also in a second aspect: independent replication and verification of experimental results in mobile robotics is, as yet, uncommon. While in sciences such as biology or physics, for instance, reported results are only taken seriously once they have been verified independently a number of times, in robotics this is not the case. Instead, papers often describe experimental results obtained in specific environment, under specific experimental conditions. These experiments therefore are "existence proofs" — the demonstration that a particular result can be achieved — but they do not state in general terms under which conditions a particular result can be obtained, nor which principles underlie the result. Existence proofs are useful, they demonstrate that something can be achieved, which is an important aspect of science, but they do not offer general principles and theories.

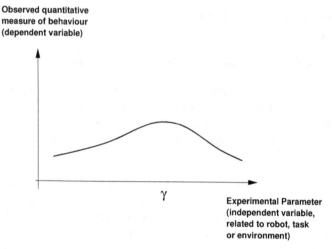

Figure 2.3. Illustration of a conceivable relationship between quantitative performance measure and experimental parameter

We argue that mobile robotics research is now at a stage where we should move on from existence proofs to a research culture that habitually includes independent replication and verification of experiments.

Theories, experimental replication and experimental verification all depend crucially on *quantitative* descriptions: quantitative descriptions are an essential element of the language of science. For these reasons this book presents several ways of describing robot-environment interaction quantitatively[1].

2.5 Robot Engineering *vs* Robot Science

Arguably, there are (at least) two independent objectives of robotics research: on the one hand, to create artefacts that are capable of carrying out useful tasks in the real world — for example industrial, service, transportation or medical robots, to name but a few, and on the other hand to obtain a theoretical understanding of the design issues involved in making those artefacts — for example sensor and actuator modelling, system identification (modelling of entire systems), or sensor, actuator and behaviour analysis. The former can be referred to as "robot engineering", the latter as "robot science". It is robot science that this book is mainly concerned with.

While robot engineering ultimately produces the "useful" artefacts, there is a lot that robot science can contribute to this process. Without theoretical understanding, any design process is largely dependent upon trial-and-error exper-

---

[1] A very informative article on quantitative measures of robot-environment interaction can be found in [Smithers, 1995].

imentation and iterative refinement. In order to design in a principled way, a hypothesis — a justified expectation — is needed to guide the design process. The hypothesis guides the investigation: results obtained are fed back into the process and brought into alignment with the theory, to lead to the next stage of the experimentation and design. The better the theory underlying the design process, the more effective and goal-oriented the design process will be. *Every process of designing technical artefacts is based on some kind of assumptions (a "theory"), even if very little is known at all about the object being designed.*

This is true for current mobile robotics research, too. When asked to design a wall-following robot, the designer will not start with an *arbitrary program*, but with a "reasonable guess", sensibly speculating on which sensors might be useful to achieve the desired behaviour, which general kind of control program will perform acceptably, *etc*. But, given our current understanding of robotics, he is unable to design the entire behaviour off-line!

Instead, mobile robotics researchers to-date are crucially dependent on trial-and-error procedures. A "reasonable prototype" has to be tested in the target environment, and refined based on observations and underlying theory ("hunch" is often the more appropriate term for such theories). Here is a practical example: to design the Roomba commercial robot floor cleaner (relying on very simple sensing, and not involving any sophisticated navigation), 30 prototypes had to be built over a period of 12 years [EXN, 2003]!

Theoretical understanding of robot-environment interaction, however, would address this issue, and support off-line design. But not only that: it would further-more allow the analysis of an observed behaviour, and the refinement of existing mechanisms, based on established theoretical principles.

The argument this book makes, therefore, is this: a better theoretical under-standing of the principles underlying a mobile robot's operation in its environ-ment — a theory — will result in more effective, rigorous and goal-oriented development methods. These, in turn, will support robot engineering, leading to robots that are better able to achieve the tasks they are designed for.

2.6 Scientific Method and Autonomous Mobile Robotics

2.6.1 Introduction

Whether mobile robotics actually is a science or an engineering discipline, it undoubtedly benefits from clear, coherent and methodical research practice, and the following discussion should be relevant to both "science" and "engineering".

The discipline of mobile robotics is interested in developing artefacts (robots) that can carry out some useful task in a real world environment. However this is attempted, be it trial-and-error, methodical research or a mixture of both, the designer will rely on some previously acquired knowledge, perhaps inadver-tently. This knowledge essentially constitutes a "theory". It is useful to analyse in more detail what the elements of this theory are, and how the theory can be improved — this is the purpose of this chapter.

2.6.2 Background: What is "Scientific Method"?

As stated earlier, the aim of this book is to open up new avenues of conducting research in mobile robotics, to move away from existence proofs and the need for iterative refinement, and to overcome the inability to design task-achieving robots off line. Before we look at some practical ways of applying scientific method to mobile robotics research, we'll look at a very broad summary of what has been understood by the term "scientific method" over the centuries. For a proper treatment of this topic, however, please see dedicated books on the subject (for example [Gillies, 1996, Harris, 1970, Gower, 1997].)

Sir Francis Bacon (1561 – 1626) first developed the theory of inductivism [Bacon, 1878], where the basic idea is this: first, a large number of observations regarding the subject under investigation is gathered. This includes "instances where a thing is present", "instances where a thing is not present", and "instances where a thing varies". The nature of the phenomenon under investigation is then determined by a process of eliminative induction. Almost mechanically, by gathering more and more information and ruling out impossible hypotheses, the truth is established. [Gillies, 1996] likens this inductive process to that of drawing a precise circle: impossible to achieve just using pen and paper, but very easy using the mechanical device of a compass. In a similar manner, scientific truths are to be discovered by the mechanical process of induction. The "problem of induction", however, is that the facts gathered can never be complete enough to fully justify the conclusions drawn, so that any hypotheses are in effect working hypotheses only, a first stab, so to speak.

The complete opposite view to Bacon's induction based on many observations is Karl Popper's argument that induction is a myth, because observation without theory is impossible [Popper, 1959, Popper, 1963, Popper, 1972]. In other words, there needs to be a theory first in order to observe, and obtaining a theory from a large body of observations alone is impossible. Simply "observing" cannot be done, the scientist needs to know what should be observed. This in turn requires the definition of a chosen task, a question, a problem — in other words: a hypothesis. Instead of inductivism, he proposed a theory of conjectures and refutations (falsificationism): the aim of scientific investigation is to refute a hypothesis, and all experimentation is geared towards that goal. If a hypothesis withstands all attempts of refutation, it is *tentatively* adopted as true, but not considered proven and true without doubt. The only truth that can be firmly established is that a theory is false, never that it is true.

How then does the scientific community accept or reject theories? Thomas Kuhn [Kuhn, 1964] differentiates between "normal science" and a situation of "scientific revolution". Normal science he describes as research firmly based on past scientific achievements or "paradigms". Paradigms here refer to theories that create avenues of enquiry, formulate questions, select methods and define relevant research areas — paradigms guide research. "Normal" scientific research aims to extend the knowledge within an existing paradigm, to match facts with theory, to articulate theory and to bring the existing theory into closer agreement with observed facts. It tends to suppress fundamental novelties that cannot be brought into agreement with existing paradigms. Normal science works within the accepted, existing paradigm, seeks to extend the knowledge the paradigm is revealing, and to "tie up loose ends" and plug gaps — Kuhn refers to this as "mopping up".

However, in the process of normal science increasingly discrepancies between fact and theory (anomalies) will be observed. There will be observations that cannot be explained at all with the existing theory, and there will be observations that appear to disagree with existing theory. These difficult cases tend to be ignored initially, but their weight and importance may increase until a point is reached at which the scientific community loses faith in the existing paradigm. A crisis has developed; it begins with a blurring of the existing paradigms, continues by the emergence of proposals for alternative paradigms, and eventually leads to a "scientific revolution", the transition form "normal" to extraordinary research. Eventually, the new paradigm is adopted by the majority of scientists and assumes the role of "normal" paradigm, and the process is repeated.

Scientific Research Methodology

As stated in the introduction, this book is no attempt to present an account of philosophy of science and its application to mobile robotics. When we refer to "scientific method", the emphasis is not on the philosophical foundations of research methodology.

Rather, it is on the procedure of conducting, evaluating and reporting research and its results; that is, the material practice of science, the "recipes". What is a good starting point for research? How do we design experiments, how do we document and assess the results? What do we adopt as a scientific research procedure within the community? These are the kinds of questions we should be able to answer before we conduct the actual research!

Forming Scientific Hypotheses

The starting point for any research is a hypothesis, a thesis. This hypothesis is a formally stated expectation about a behaviour that defines the purpose and the goals of a study; it therefore defines, explains and guides the research. Without a clear hypothesis in the beginning, it is virtually impossible to conduct good research, as it is virtually impossible to present results in a coherent and convincing way. The hypothesis, the question, is the foundation upon which the scientific argument is built. Obviously, an ambiguous question will result in an ambiguous answer, which is why the hypothesis is the most fundamental stage of scientific working.

To formulate the hypothesis clearly, it is useful to consider the following points (see also [Paul and Elder, 2004]):

1. What is the question addressed?
   - State it precisely
   - Can it be broken down into sub questions?
   - Is there one right answer to the question? Does it require reasoning from more than one point of view? Is it a matter of opinion?
2. What assumptions are you making?
   - Identify all assumptions clearly
   - Are they justifiable?
   - Do these assumptions affect the impartiality of your research?
   - Identify key concepts and ideas that shape the research. Are they reasonable?
3. Formulate a hypothesis
   - Is this hypothesis testable and falsifiable?
   - What outcome do you expect?
   - What would be the implications of the different possible outcomes of your experiment (*i.e.* is the question actually worth asking)?
   - Experimental design
4. Which experimental setup is suitable to investigate the question/hypothesis?
   - How is experimental data going to be collected?
   - How is experimental data going to be evaluated?
   - How much data is needed?

Hypotheses can be causal hypotheses, hypothesising about the causes of a behaviour, or descriptive, describing a behaviour in terms of its characteristics or the situation in which it occurs. Causal reasoning and causal models are very common in science, and guide experimental design, hypothesis formation and the formation of theories. Causal models guide scientific thinking so strongly that on occasions scientists even override the statistical information they receive, in favour of a causal model [Dunbar, 2003] (referred to as "confirmation bias" — "cold fusion" being a prominent example). In other words: the hypotheses guiding research can be so dominant that the scientist tries to generate results that confirms his initial hypothesis, rather than aiming to disprove a hypothesis (which is, according to Popper, what he should be doing!) [Klayman and Ha, 1987] — the tendency of trying to *confirm* a hypothesis, rather than refute it, is difficult to overcome. The temptation to conduct experiments that produce results predicted by the current hypothesis is very strong!

Popper argued that (due to the infinity of the universe) scientific hypotheses can *never* be verified (*i.e.* proven to be true) nor the probability of their veracity established, but that they can only be falsified, *i.e.* shown to be incorrect. He further argued that the most fundamental requirement for any scientific hypothesis must therefore be that the theory is open to tests and open to revision. In other words: it must be testable, and it must be falsifiable. If either of these conditions isn't met, the hypothesis will not support scientific investigation.

Popper was aware that it is possible to evade falsification by adopting "saving stratagems" (*e.g.* by modifying testability of a hypothesis), and therefore introduced the supreme rule that "the other rules of scientific procedure must be designed in such a way that they do not protect any statement in science from falsification" [Popper, 1959, p.54].

"The aim of science is to find satisfactory explanations, of whatever strikes us as being in need of explanation" [Popper, 1972, p. 191] — the hypothesis underlying the research ultimately defines the degree to which an explanation is satisfactory or not.

There are further criteria that distinguish "good" hypotheses from "bad" ones. Popper and Kuhn identify explanatory depth as a crucial aspect — which paradigm explains more phenomena? —, but increased verisimilitude is equally identified by Popper as an objective for forming hypotheses. In a survey article, summarising the views put forward by Kuhn, Lakatos and Laudan, [Nola and Sankey, 2000] state that "Scientists prefer a theory that

- Can solve some of the empirical difficulties confronting its rivals
- Can turn apparent counter-examples into solved problems
- Can solve problems it was not intended to solve
- Can solve problems not solved by its predecessors
- Can solve all problems solved by its predecessors, plus some new problems
- Can solve the largest number of important empirical problems while generating the fewest important anomalies and conceptual difficulties"

Hypotheses must be precise, rational (that is, possibly true and in agreement with what is already known) and parsimonious (that is, as simple as possible — but not simpler. William of Occam's razor — "entities are not to be multiplied beyond necessity" — is one expression of this principle). In summary, the hallmarks of a "good" scientific paradigm — which must be testable and falsifiable — are explanatory power, clarity and coherence.

How can scientific hypotheses be obtained? The most common sources are:

- Opinions, observations and experiences
- Existing research
- Theories
- Models

Karl Popper argued that scientific hypotheses are the product of brilliant creative thinking by the scientist (he refers to this as "creative intuition").

2.6.3 Experimental Design and Procedure

Experimental Design

Experimental design — the experimental procedure used, the observation mechanisms and the way results are interpreted — is the centre of any scientific investigation, and care is necessary when designing experiments. Is the chosen design suitable for investigating the hypothesis I am interested in? Is there a better way of achieving my objectives? Is the design feasible in practice, or does it offer insurmountable practical problems?

One of the most common types of scientific experiments aim to determine a relationship between two variables: one that is controlled by the experimenter (the independent variable, IV), and one that is dependent on it (the dependent variable, DV). The most common aim of experimentation is to establish how the DV changes in relation to the IV.

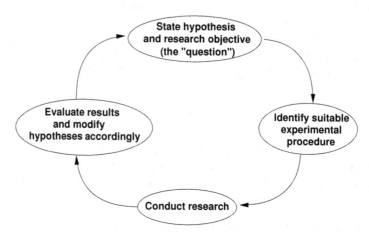

Figure 2.4. Experimental procedure

There are a number of aspects to be considered when designing an experiment (see also Figure 2.4):

- What is the question being asked? What is the hypothesis? *Every* scientific investigation is driven by the underlying question it has set out to answer. If this question is not formulated clearly, or even not formulated at all, the resulting research will be haphazard, ill focused without clear aim. Good research needs a clearly formulated objective!
- Sensitivity of the experiment. Is the experimental design sensitive enough to detect any causal relationship between DV and IV? Is it perhaps too sensitive, and will therefore amplify noise?

- Are there any confounding effects that introduce errors that hide any true effects (see below for a discussion of this point)?
- Which predictions can be made about the outcome of the experiment? Are there expectations, perhaps based on some prior understanding of the problem, that can be used to assess the eventual outcome of the experiment?
  Predictions are important, they serve as a sanity check, helping us to identify results that are highly unlikely, and to detect possible errors in the experimental design, procedure and evaluation.
- Observation. How is the experiment going to be observed, how are results going to be recorded? It is important to strive for consistency here. Similar experiments should result in similar results, if they don't, one has to check the experimental design again for possible error sources and procedural mistakes.
- Analysis and interpretation. How are the results going to be analysed? Merely describing experimental results in words is a possibility, but there are stronger tools available for analysis. Chapter 3 covers a whole range of statistical methods that can be used to detect "significant" effects.
  A very important method used in analysis and interpretation is that of *creating a baseline*. The baseline serves as the fundamental data against which one compares the results obtained in the experiment.
  For example, in work concerned with predictions (of, for example, robot trajectories, temperatures in your home town or stock market developments) one very often compares a prediction method against the baseline of predicting the mean. Predicting that a future signal value will be the mean of past values is a very reasonable prediction, which tends to minimise prediction error. If a prediction method is claimed to be "good", it ought to outperform this simple predictor — something that can be established by the methods described in Chapter 3.
- Often it is useful to conduct a pilot study first, in order to minimise the experimental effort. A pilot study investigates the underlying question in a "broad shot" manner, eliminating certain possibilities, making others more likely, while using simplified and coarser experimental procedures than the eventual final study.

Traps and Countermeasures

Traps

There are a number of known traps to avoid [Barnard et al., 1993]:

1. Confounding effects. If the phenomenon we are interested in is closely correlated with some other effect that is of no interest, special care has to be taken to design the experiment in such a way that only the factor of interest is investigated.

For example, we might be interested in measuring whether the obstacle avoidance movements of a learning mobile robot become more "efficient", smoother, with time. We might find a positive correlation, and conclude that our learning algorithm results in ever smoother movement. But unless we design our experiment carefully, we cannot be sure that the increasingly smooth movement is not the result of decreasing battery charge, resulting in a sluggish response of the robot's motors!

2. Floor and ceiling effects. It is possible that the experimental design is either too demanding or too simple to highlight relevant phenomena.

    For example, we might be interested to investigate whether one service robot performs better than another. If we compare both robots in too simple an environment, they might not show any difference whatsoever (floor effect). On the other hand, if we choose a very complicated environment, neither robot may perform satisfactorily (ceiling effect). Obviously, in order to highlight any differences between the two robots, just the right type of environment complexity is needed.

3. Pseudo-replication (non-independence). The more costly (in terms of time or resources) an experiment, the greater the risk to produce data that is not independent, so-called pseudo-replication. Pseudo-replication means that the errors of our measurements are not unique to each measurement, *i.e.* not independent.

    For example, we might want to measure what effect the colour of objects has on a robot's ability to detect them with its camera system. We could take three different objects, say, and let the robot detect each of these objects ten times. This does not, however, result in thirty independent measurements! We really only have three independent measurements in this case, and need to collapse the ten observations for each object into one value, before we proceed with an analysis of the results.

4. Constant errors, that is systematic errors (biases) can mask true effects, and need to be avoided.

5. "The conspiracy of goodwill" (Peter Medawar). In designing our experiments we need to take great care to retain objectivity. It is very easy to have a particular desired outcome of our experiments in mind, and to research selectively to attain that outcome!

Countermeasures

There are a range of countermeasures that can be taken to avoid the pitfalls just mentioned.

First of all, it is good practice to include *controls* in the experimental design. Such controls can take the form of experiments within the chosen experimental setup whose results are known. Say, for example, an exploration robot is designed to detect certain objects (*e.g.* rocks) in some remote area (*e.g.* Antarctica). The usual procedure, the control, is to test the robot and its ability to detect the objects in a laboratory environment, where the robot's detection ability can be observed and measured.

A second, very commonly used and very effective method to counteract pitfalls of scientific investigation is to work in groups, and to seek independent verification and confirmation of one's experimental setup, experimental procedure, results and their interpretation. Usually hypotheses, setups and interpretations benefit from independent scrutiny!

Constant errors can be avoided by *counterbalancing and randomisation*. Counterbalancing stands for an experimental procedure in which each arrangement of variables under investigation is used an equal number of times. If, for instance, two different robot controllers A and B are to be tested in the same environment, a counterbalanced experimental design would mean that A and B are used first and second respectively for an equal number of time. This would counterbalance constant errors introduced by wear and tear, such as decreasing battery charge.

Another method of dealing with constant errors is that of *randomisation*, by which we mean counterbalancing by chance: the arrangement of variables is determined randomly.

Counterbalancing can only be used if there is no interaction between the counterbalanced variables. If, for example, program B of the above example modified the environment, for instance by rearranging objects in the environment, it does matter in which sequence programs A and B are tested. Counterbalancing would not work in this case.

Dealing with the "conspiracy of goodwill" is relatively easy: a "blind" experimental arrangement will achieve that. *Blind experimentation* means that the experimenter is unaware of the state of the independent variable, and therefore has to log and interpret resulting experimental data at face value, rather than inadvertently putting a slant on the interpretation.

Best known for trials in medicine, where the scientific question is whether a particular drug is effective or not (independent of the patient's and the doctor's knowledge of which drug or placebo is being administered), blind experimentation actually also has a place in robotics. The temptation to interpret results in favour or one's own control program in comparison with a baseline control program is always there! If the experimenter is unaware of which program is currently being run, he cannot possibly log and interpret the data in a biased way!

2.7 Tools Used in this Book

2.7.1 Scilab

In some chapters of this book we have included numerical examples of methods and algorithms discussed in the text. We have used the mathematical programming package `Scilab` [Scilab Consortium, 2004] to illustrate the examples, and included listings of some programs. Many figures in this book were generated using `Scilab`.

`Scilab` is a powerful mathematical programming language, which, as a bonus, has the advantage that it is free for personal use. However, the examples given in this book require few changes to run on other mathematical programming languages, such as for example `Matlab`.

2.8 Summary: The Contrast Between Experimental Mobile Robotics and Scientific Mobile Robotics

In summary, the contrast between mobile robotics as an experimental discipline and mobile robotics as a scientific discipline can be described like this:

- Experimental design and procedure is guided by a testable, falsifiable hypothesis, rather than based on the researcher's personal experience (a "hunch")
- Experimental design and procedure are "question-driven", rather than "application-driven"
- Results are measured and reported quantitatively, rather than qualitatively
- Experimental results are replicated and verified independently (for example by other research groups), rather than presented as stand-alone existence proofs

The following sections of this book will look at how these objectives can be achieved. How can the performance of a mobile robot be assessed, and compared with that of an alternative control program? How can robot-environment interaction be described *quantitatively*? How can testable hypotheses be formulated? How can robot-environment interaction be modelled and simulated accurately? These are the questions that we will investigate now.

3

Statistical Tools for Describing Experimental Data

Summary. Statistical descriptions of experimental data are one of the simplest methods of describing *quantitatively* what a robot does. This chapter presents statistical methods that are useful when analysing experimental data generated by an agent such as a robot, and gives examples of applications in robotics.

3.1 Introduction

This chapter looks at a wide range of statistical techniques that can be used to analyse, describe or quantify robot behaviour. Many of these procedures are taken from the life sciences, where statistical analysis and comparison of behaviour is well established.

In any experimental science, be it biology, psychology, medicine or robotics, to name but a few, we typically perform experiments designed to test our hypothesis. The experiment is observed, data describing the relevant aspects of the experiments is logged, and subsequently analysed. Once such data is logged, one typically wants to answer some of these questions:

- Is there a statistically significant correlation between input and output variables?
- Is there a statistically significant difference between the experimental results obtained and some "baseline" (either another method to achieve the same task, or a well established mechanism that is well understood)?
- Alternatively: could the experimental results be explained by random events? Are they a fluke?

"Statistically significant" here is a precisely defined technical term, meaning that the outcome of an experiment differs from the "null hypothesis"[1] by more

---

[1] The hypothesis that the observed outcome of an experiment is due to chance alone, and not due to a systematic cause.

than what could be attributed to random fluctuations. "Significance" is discussed later in this chapter.

Statistical tests involve i) determining what kind of data is to be analysed, ii) determining what kind of question (null hypothesis) is being asked, iii) selecting an appropriate test, and iv) performing the analysis.

This chapter first introduces the kind of data that might be analysed (normally distributed or not normally distributed), then looks at methods to determine whether two samples are drawn from the same underlying distribution or not (*i.e.* whether they are significantly different from each other or not), then looks at tests that determine whether there exists a significant trend that could describe the relationship between two variables, and finally presents methods that are suitable for analysing categorical data (basically, data that is not numerical, but based on categories). Table 3.1 gives an overview of the procedures introduced in this chapter.

Table 3.1. Statistical methods discussed in this chapter

	Are samples from the same distribution?	Tests for a trend
Data is normally distributed	Mean and std. dev (Section 3.2) t-test (Section 3.3.4) param. ANOVA (Section 3.3.6) U-statistic (Section 3.4.2) Wilcoxon test (Section 3.4.3) non-param. ANOVA (Section 3.4.4)	Linear regression (Section 3.6.1) Correlation analysis (Section 3.6) Spearman rank correlation (Section 3.7.1)
Data is not normally distributed	Median (Section 3.4.1) U-statistic (Section 3.4.2) Wilcoxon test (Section 3.4.3) non-param. ANOVA (Section 3.4.4)	Spearman rank correlation (Section 3.7.1)
Categorical data		χ^2 (Section 3.8.1) Cramer's V (Section 3.8.2) Entropy (Section 3.8.3)

3.2 The Normal Distribution

The most common distribution of values, for example obtained by measuring some physical entity, is the Gaussian distribution. Because it is the usually occurring distribution, it is often referred to as the "normal distribution".

Mean, Standard Deviation and Standard Error

The Gaussian or normal distribution can be completely described by two parameters, mean μ and standard deviation σ — hence the term "parametric" for distributions like this. For the normal distribution, 68.3% of all measurements x_i lie in the interval $\mu \pm \sigma$, 95.4% of all x_i lie in the interval $\mu \pm 2\sigma$, and 99.7% of all x_i in the interval $\mu \pm 3\sigma$.

In a Gaussian (normal) distribution, values are centred around the "expected value", the "mean" μ. The width of the bell-shaped curve is determined by the so-called "standard deviation" σ — more about that below.

The probability density of this normal distribution, $p(x)$, is shown in Figure 3.1 and defined by Equation 3.1.

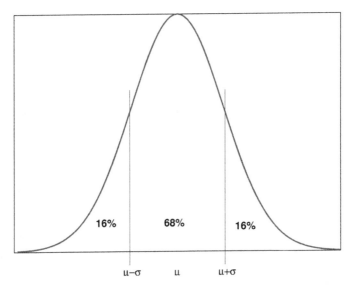

Figure 3.1. Gaussian or normal distribution. Values are centred around the mean μ, with outliers becoming less frequent the further from the mean they are; 68.3% of all measurements are in the interval $\mu \pm \sigma$

$$p(x) = \frac{1}{\sigma\sqrt{2\pi}}e^{\frac{-(x-\mu)^2}{2\sigma^2}} \tag{3.1}$$

The expected value of our measurement – the mean μ – is defined by Equation 3.2, the standard deviation σ is given by Equation 3.3[2]:

$$\mu = \frac{1}{n}\sum_{i=1}^{n} x_i \tag{3.2}$$

where x_i is one individual measurement from the series of measurements, and n is the total number of measurements.

---

[2] Equations 3.2 and 3.3 are approximations that are used in practice. Strictly speaking, $\mu = \lim_{n\to\infty} \frac{1}{n}\sum_{i=1}^{n} x_i$ and $\sigma = \lim_{n\to\infty} \sqrt{\frac{1}{n-1}\sum_{i=1}^{n}(x_i - \mu)^2}$.

$$\sigma = \sqrt{\frac{1}{n-1}\sum_{i=1}^{n}(x_i - \mu)^2}$$

(3.3)

Mean and standard deviation fully describe the normal distribution, and contain information about the precision of our measurements x_i, in that σ indicates what percentage of measurements will lie in a specific interval $\mu \pm k\sigma$. For $k = 1$, for instance, this percentage is 68.3%.

As the individual measurements x_i, so the mean itself is also subject to error. We can determine the mean error of the mean, $\overline{\sigma}$, as

$$\overline{\sigma} = \frac{\sigma}{\sqrt{n}}$$

(3.4)

This so-called *standard error* $\overline{\sigma}$ is a measure for the uncertainty of the mean, and $\mu \pm \overline{\sigma}$ denotes that interval within which the true mean lies with a certainty of 68.3%, $\mu \pm 2\overline{\sigma}$ the interval within the true mean lies with a certainty of 95.4% and $\mu \pm 3\overline{\sigma}$ is the 99.7% confidence interval. When stating means in the literature, or plotting them in bar graphs, it is common to report them as $\mu \pm \overline{\sigma}$.

Reducing the Measuring Error

As the number of measurements increases the standard error $\overline{\sigma}$ decreases. This means that the deviation of the mean μ from the true mean also decreases. However, as the standard error is proportional to the mean error of the individual measurement (Equation 3.3) and inversely proportional to $\sqrt{n}$, it is not useful to increase the number of measurements arbitrarily to reduce uncertainty. If we want to reduce the measuring error, it is better to increase the measuring precision.

The Standard Normal Distribution

The normal distribution plays an important role in evaluating the outcome of many statistical tests. Usually, the standard normal distribution of $\mu = 0$ and $\sigma = 1$ is used for these comparisons; this distribution is given in Table 3.2, which gives the area underneath the normal distribution curve in the interval $\mu + z$ (see also Figure 3.2).

Any normal distribution with means other than zero and standard deviations other than one can be transformed into the standard normal distribution through Equation 3.5:

$$z(x) = \frac{x - \mu}{\sigma}$$

(3.5)

Table 3.2. Standard normal distribution table, giving the area from μ to $\mu + z$ under the standard normal distribution of $\mu = 0$ and $\sigma = 1$

z	.00	.02	.02	.03	.04	.05	.06	.07	.08	.09
0.0	0.000	0.004	0.008	0.012	0.016	0.020	0.024	0.028	0.032	0.036
0.1	0.040	0.044	0.048	0.052	0.056	0.060	0.064	0.067	0.071	0.075
0.2	0.079	0.083	0.087	0.091	0.095	0.099	0.103	0.106	0.110	0.114
0.3	0.118	0.122	0.126	0.129	0.133	0.137	0.141	0.144	0.148	0.152
0.4	0.155	0.159	0.163	0.166	0.170	0.174	0.177	0.181	0.184	0.188
0.5	0.191	0.195	0.198	0.202	0.205	0.209	0.212	0.216	0.219	0.222
0.6	0.226	0.229	0.232	0.236	0.239	0.242	0.245	0.249	0.252	0.255
0.7	0.258	0.261	0.264	0.267	0.270	0.273	0.276	0.279	0.282	0.285
0.8	0.288	0.291	0.294	0.297	0.300	0.302	0.305	0.308	0.311	0.313
0.9	0.316	0.319	0.321	0.324	0.326	0.329	0.331	0.334	0.336	0.339
1.0	0.341	0.344	0.346	0.348	0.351	0.353	0.355	0.358	0.360	0.362
1.1	0.364	0.367	0.369	0.371	0.373	0.375	0.377	0.379	0.381	0.383
1.2	0.385	0.387	0.389	0.391	0.393	0.394	0.396	0.398	0.400	0.401
1.3	0.403	0.405	0.407	0.408	0.410	0.411	0.413	0.415	0.416	0.418
1.4	0.419	0.421	0.422	0.424	0.425	0.426	0.428	0.429	0.431	0.432
1.5	0.433	0.434	0.436	0.437	0.438	0.439	0.441	0.442	0.443	0.444
1.6	0.445	0.446	0.447	0.448	0.449	0.451	0.452	0.453	0.454	0.454
1.7	0.455	0.456	0.457	0.458	0.459	0.460	0.461	0.462	0.462	0.463
1.8	0.464	0.465	0.466	0.466	0.467	0.468	0.469	0.469	0.470	0.471
1.9	0.471	0.472	0.473	0.473	0.474	0.474	0.475	0.476	0.476	0.477
2.0	0.477	0.478	0.478	0.479	0.479	0.480	0.480	0.481	0.481	0.482
2.1	0.482	0.483	0.483	0.483	0.484	0.484	0.485	0.485	0.485	0.486
2.2	0.486	0.486	0.487	0.487	0.487	0.488	0.488	0.488	0.489	0.489
2.3	0.489	0.490	0.490	0.490	0.490	0.491	0.491	0.491	0.491	0.492
2.4	0.492	0.492	0.492	0.492	0.493	0.493	0.493	0.493	0.493	0.494
2.5	0.494	0.494	0.494	0.494	0.494	0.495	0.495	0.495	0.495	0.495
2.6	0.495	0.495	0.496	0.496	0.496	0.496	0.496	0.496	0.496	0.496
2.7	0.497	0.497	0.497	0.497	0.497	0.497	0.497	0.497	0.497	0.497
2.8	0.497	0.498	0.498	0.498	0.498	0.498	0.498	0.498	0.498	0.498
2.9	0.498	0.498	0.498	0.498	0.498	0.498	0.498	0.499	0.499	0.499
3.0	0.499	0.499	0.499	0.499	0.499	0.499	0.499	0.499	0.499	0.499

with x being a value of the original normal distribution that is to be transformed into the standard normal distribution.

Table 3.2 can then be used to determine areas underneath the distribution curve. For instance, the area between μ and 1 is, according to the Table, 0.341. Because the area underneath the entire curve of the standard normal distribution is one, this means that 34% of all data points lie between $\mu + 1\sigma$, or 68% within the interval of $\mu \pm \sigma$.

3.3 Parametric Methods to Compare Samples

3.3.1 General Considerations (Significance Levels)

When conducting statistical tests, one typically determines some test statistic S, and evaluates whether S lies within an acceptance interval $\mu \pm z\sigma$. z is determined by the significance level specified by the user. This significance level indicates the combined area underneath the lower and upper tail of the distri-

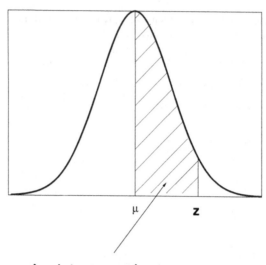

Area between μ **and z**

Figure 3.2. Table 3.2 gives the area between the mean μ of a normal distribution and z, in per cent of the whole area underneath the curve

bution (shaded areas in Figure 3.3), and therefore determines the width of the acceptance interval.

If the sample statistic falls outside the acceptance interval, the null hypothesis[3] is rejected; otherwise it is accepted (which simply means there is no valid reason to reject it, but nevertheless it needn't be true).

Determining a Suitable Significance Level

The acceptance interval obviously influences whether a null hypothesis is accepted or rejected: the wider the acceptance interval (*i.e.* the lower the significance level), the greater the chance that a null hypothesis is accepted.

In selecting the appropriate significance level, one has to make a decision as to whether in the particular situation it is preferable to potentially reject a true null hypothesis (type I error — high significance level), or to accept a false null hypothesis (type II error — small significance level). These errors are interdependent; one can only choose one of these to be small at the cost of making the other bigger. The appropriate significance level is usually determined by taking the cost into account that either error would incur. In scientific research, it is common to use the 5% significance level.

---

[3] For example, that there is no significant difference between two distributions.

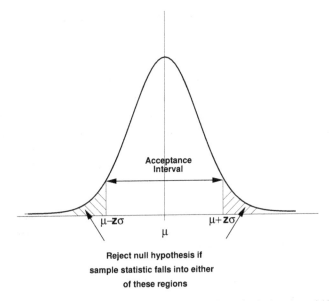

Figure 3.3. Illustration of significance levels. The null hypothesis is rejected if the sample statistic falls outside the acceptance interval

Example

A normally distributed sample statistic with $\mu_K = 50$ and $\sigma_K = 6$ has been determined to evaluate the outcome of an experiment. What is the acceptance interval at the 5% and the 10% significance level?

From Table 3.1 we see that $z = 1.96$ for an area of 47.5% ($2 \times 47.5\% = 95\%$). The acceptance interval for a significance level of 5% is given as $\mu_K \pm z\sigma_K = 50 \pm 1.96 \times 6$ (*i.e.* 38.2 to 61.8).

At the 10% significance level, we would obtain a z of 1.65 from Table 3.1, resulting in an acceptance interval of $50 \pm 1.65 \times 6 = 40.1$ to 59.9.

Therefore, if for instance a sample statistic had a value of 61, we would accept the null hypothesis at the 5% significance level, but reject it at the 10% significance level.

The fundamental assumptions in all parametric methods is that the data being analysed follows a normal distribution — parametric methods can only be applied to data with known distributions. Because of this restriction, parametric methods are less "robust" than non-parametric methods, as they will give erroneous results if applied to data that does not follow the assumed distribution (which is usually the normal distribution). On the other hand, they are better able to reject a null hypothesis than non-parametric tests, a property that is usually referred to as being more "powerful".

3.3.2 Determining Whether or not a Distribution is Normal

As stated above, all parametric tests assume that the distribution underlying the data is (approximately) normal. Often, however, this is not known *a priori*, and if we want to use parametric tests it is necessary to investigate whether the a normal distribution describes the data well or not.

For very small sample sizes ($n < 10$) it is usually not possible to determine whether the data points follow a normal distribution or not. Because tests designed for normally distributed data will give erroneous results if the data follows some other distribution, it is advisable to use non-parametric tests, rather than parametric ones, because non-parametric tests make no assumptions about underlying distributions.

For sample sizes $n > 10$, the simplest method to see whether the underlying distribution is normal or not is to use the human eye and to compare the histograms of the data in question with that of a normal distribution. Figure 3.4 shows the histogram of the obstacle avoidance behaviour of a mobile robot shown in Figure 4.12, in comparison with the histogram of a normal distribution. Even with the naked eye it is obvious that that distribution is more bimodal (two distinct peaks) rather than normal.

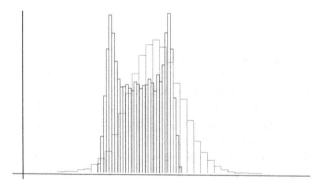

Figure 3.4. Histogram of robot obstacle avoidance in comparison with the histogram of a normal distribution

A more refined graphical technique is to plot the data on so-called normal probability paper, whose coordinates are chosen such that normally-distributed data appears as a line. One then tests visually whether the data actually fits a line well or not.

In Scilab, the equivalent of plotting on normal probability paper can be achieved by the following two commands (note that the first command stretches over three lines):

```
plot2d([0 sum(abs(a)<0.25) sum(abs(a)<0.5) sum(abs(a)<1)
sum(abs(a)<2) sum(abs(a)<3)  length(a)],
[0 20 38.3 68.2 95.4 99.7 100])

xpoly([0 length(a)],[0 100])
```

where a is the sample to be analysed. Figure 3.5 shows plots of a normally distributed random number sequence and a uniformly distributed random number sequence, using this method.

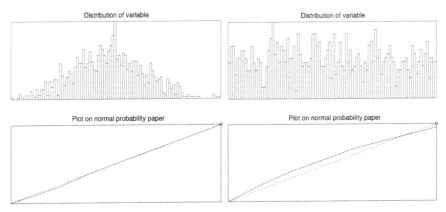

Figure 3.5. Histogram and plot on normal probability paper of a normal distribution (*left*) and a uniform distribution (*right*). For the normal distribution, a plot on normal probability paper results in a straight line, for any other distribution this is not the case

3.3.3 Dependent or Independent Samples?

Some tests are given for "independent" or "dependent" samples. "Dependent samples" refers to pairs of data, for instance measurements of the same robot's performance before and after some modification. As a rule of thumb, samples are dependent if you can't shuffle the two sets of data without losing information: if you shuffle paired data such as before-after type data, you will lose vital information, hence such data is dependent. On the other hand, shuffling the data of a sample of one robot brand, and that of another robot brand, will not lose information: such data is independent.

Another way of describing paired data is that this is data where one isn't interested in the statistical properties of either sample, but in the *difference* between the samples.

3.3.4 Comparing Two Samples: The t-Test for Independent Samples

It is often useful to have some measure of performance of a particular algorithm, control mechanism, *etc*. If, for example, two different control programs produced two different means of a particular result, it is necessary to decide whether there is a significant difference between these two means, in order to determine whether one of the two programs produces better results than the other.

The t-test is used to compare two means μ_1 and μ_2 from normally distributed values, whose standard deviations are (roughly) equal. The null hypothesis H_0 that is to be tested is $\mu_1 = \mu_2$.

The test statistic used in the t-test is the value T given in Equation 3.6; if T lies outside the acceptance interval, the null hypothesis H_0 is rejected:

$$T = \frac{\mu_1 - \mu_2}{\sqrt{(n_1 - 1)\sigma_1^2 + (n_2 - 1)\sigma_2^2}} \sqrt{\frac{n_1 n_2 (n_1 + n_2 - 2)}{n_1 + n_2}} \qquad (3.6)$$

with n_1 and n_2 being the number of data points in experiment 1 and experiment 2 respectively, μ_1 and σ_1 mean and standard deviation of experiment 1, and μ_2 and σ_2 mean and standard deviation of experiment 2.

The test is conducted as follows: the critical value of t_{crit} is determined from Table 3.3, with $k = n_1 + n_2 - 2$ being the number of degrees of freedom (this Table gives the critical values for a so-called "two-tailed test", testing whether one mean is either significantly larger *or smaller* than the other — in other words, testing whether the test statistic T is in *either* tail of the distribution). If the inequality $|T| > t_{crit}$ holds, the null hypothesis H_0 is rejected, meaning that the two means differ significantly. The probability that the outcome of the t-test is wrong, *i.e.* that it indicates significance when there is none and *vice versa*, is dependent on the significance level chosen (see also Section 3.3.1), *i.e.* the critical value t_{crit} selected. It is most common to take take the values for the 5% significance level (p=0.05), but depending on whether type I or type II errors are less desirable a different significant level may be more appropriate.

Table 3.3. Critical values t_{crit} for the two-tailed t-test, at 2%, 5% and 10% significance levels, for k degrees of freedom

k	1	2	3	4	5	6	7	8
$t_{0.02}$	31.821	6.965	4.541	3.747	3.365	3.143	2.998	2.896
$t_{0.05}$	12.706	4.303	3.182	2.776	2.571	2.447	2.365	2.306
$t_{0.10}$	6.314	2.920	2.353	2.132	2.015	1.943	1.895	1.860

k	9	10	14	16	18	20	30	∞
$t_{0.02}$	2.821	2.764	2.624	2.583	2.552	2.528	2.457	2.326
$t_{0.05}$	2.262	2.228	2.145	2.12	2.101	2.086	2.042	1.960
$t_{0.10}$	1.833	1.812	1.761	1.746	1.734	1.725	1.697	1.645

Instead of using tables such as Table 3.3, the t-distribution can be easily computed, using a range of commercially available software packages. In Scilab, for instance, t_{crit} can be computed by typing

```
[T]=cdft("T",k,P,Q)
```

where k is the number of degrees of freedom, Q half the significance level desired (because we are using a two-tailed test), and P=1-Q.

To determine, for instance, t_{crit} for k=19 at a significance level of 5%, one gets

```
[T]=cdft("T",19,0.975,0.025)
T   =   2.0930241
```

t-Test Example: Dead End Escape

A robot control program is written to enable robots to withdraw from dead ends. In a first version of the program, the robot takes the following time in seconds to escape from a dead end:

Exp1=(10.2, 9.5, 9.7, 12.1, 8.7, 10.3, 9.7, 11.1, 11.7, 9.1).

After the program has been improved, a second set of experiments yields these results: Exp2=(9.6, 10.1, 8.2, 7.5, 9.3, 8.4).

These results are shown in Figure 3.6. Do they indicate that the second program performs significantly better?

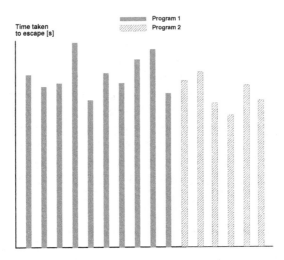

Figure 3.6. Results obtained by two different dead-end-escape programs

Answer: Assuming that the outcome of the experiments has a normal (Gaussian) distribution[4], we can apply the t-test to answer this question. $\mu_1 = 10.21, \sigma_1 = 1.112, \mu_2 = 8.85, \sigma_2 = 0.977$.

Applying Equation 3.6 yields

$$T = \frac{10.21 - 8.85}{\sqrt{(10-1)1.112^2 + (6-1)0.997^2}} \sqrt{\frac{10 \times 6(10+6-2)}{10+6}} = 2.456 \ .$$

As $k = 10 + 6 - 2$, $t_\alpha = 2.145$ (from Table 3.3). The inequality $|2.456| > 2.145$ holds, the null hypothesis H_0 (i.e. $\mu_1 = \mu_2$) is rejected, which means that the second program performs significantly better than the first one, the probability for this statement to be erroneous is 0.05.

3.3.5 The t-Test for Dependent Samples

If two tests were conducted with the same subject in each test, for example the same robot executing task x and task y, the data obtained from such an experiment is paired, or "dependent". In this case, the following t-test is applicable.

To conduct the test, the t_d-statistic is computed as indicated in Equation 3.7:

$$t_d = \frac{\mu_x - \mu_y}{S_D}; \tag{3.7}$$

with μ_x and μ_y being the means of the two measurements, and S_D the standard error of the difference, given in Equation 3.8:

$$S_D = \frac{1}{\sqrt{N}} \sqrt{\frac{\sum D^2 - \frac{(\sum D)^2}{N}}{N-1}} \tag{3.8}$$

$D = X - Y$ is the difference between the score achieved in task x and the corresponding score in task y, and N the total number of paired scores. "Score" here refers to the measurement by which we assess the system under investigation — e.g. in the following example the speed attained by a mobile robot.

To determine significance, one compares t_d against the critical t_{crit} values given in Table 3.3 for $k = N - 1$ degrees of freedom, at the desired significance level. If $|t_d| > t_{crit}$, the null hypothesis (that there is no significant difference between the two samples) is rejected.

t-Test for Dependent Samples Example: Rubber Wheels *vs* Plastic Wheels

The wheels of a factory transportation robot are changed from hard plastic wheels to softer rubber wheels, which give a better grip. The robot is sent along

---

[4] Sample sizes are really too small to make this assumption in this case, but have been chosen so small to make the example clearer.

identical tracks on different surfaces, first with the plastic wheels attached, then with the rubber wheels. For each type of wheel 12 experiments are conducted, and the average speed attained by the robot in each experiment is logged. We would like to know whether there is a significant difference in robot speed when using rubber wheels, as opposed to using plastic wheels. As the experiment is carried out under identical environmental conditions, and using the same robot, this is a paired test, the samples are dependent.

The test results obtained are shown in Table 3.4, together with some calculations useful to this test.

Table 3.4. Speeds attained by the same robot, using plastic or rubber wheels respectively, executing the same task for each pair of data

													Sum	μ
Rubber	58.	31.	82.	87.	40.	42.	71.	45.	35.	23.	13.	24.		45.9
Plastic	59.	146.	111.	111.	75.	60.	57.	81.	13.	59.	106.	37.		76.3
D	- 1.	- 115.	- 29.	- 24.	- 35.	- 18.	14.	- 36.	22.	- 36.	- 93.	- 13.	-364	-30.3
D^2	1	13225	841	576	1225	324	196	1296.	484.	1296.	8649.	169	28282	

From Equation 3.8 follows $S_D = \frac{1}{\sqrt{12}}\sqrt{\frac{28282-\frac{(-364)^2}{12}}{12-1}} = 11.4$. With $\mu_{rubber} = 45.9$ and $\mu_{plastic} = 76.3$ Equation 3.7 then yields $t_d = \frac{45.9-76.3}{11.4} = -2.66$.

With $k = N - 1 = 11$ degrees of freedom we can see from Table 3.3 that for the 5% significance level $t_{crit} \approx 2.2$. As $|t_d| > t_{crit}$, we reject the null hypothesis: there is a statistically significant difference between the speeds achieved, using plastic or rubber tyres. The robot is significantly faster, using rubber wheels.

In Scilab, we can determine the significance level exactly, using the Scilab command [P,Q]=cdft("PQ",T,Df):

```
[Q P]=cdft("PQ",2.66,11)
P   =


    0.0110934
```

P is half the significance level of this two-tailed test, so that the result is significant with p=0.022.

An alternative method of arriving at the same result is to compute an acceptance interval, using the t-statistic, and to determine whether the sample statistic in question lies within the acceptance interval or not.

In this case, our null hypothesis is $H_0 : \Delta_0 = \mu_{rubber} - \mu_{plastic} = 0$. The mean difference in speed is $\mu_{diff} = 30.3$, the standard deviation $\sigma_{diff} = 39.6$, and the standard error $\overline{\sigma}_{diff} = \frac{\sigma_{diff}}{\sqrt{n}} = 8.8$. For a two-tailed t-test at a significance level of 5% we get t=2.2 for $k = 12 - 1$ degrees of freedom from Table 3.3. The acceptance interval therefore is $\Delta_0 \pm t \times \overline{\sigma}_{diff}$, *i.e.* in this example $0 \pm 2.2 \times 8.8$ (-19.3 to 19.3). The mean difference of 30.3 is outside that acceptance interval, therefore the change in speed is significant at the 5% significance level.

3.3.6 Comparing More than Two Samples: Parametric Analysis of Variance

In Section 3.3.4 we discussed how significant difference between two means of parametric data can be determined. We might, for instance, have run a mobile robot on two different kinds of floor surface, and measured the average speed of the robot for each "treatment", wanting to know whether the robot speed is significantly different between the two cases.

But if we now introduce a third treatment, *i.e.* a third kind of floor covering on which we let the robot run, the t-test introduced in Section 3.3.4 can no longer be used. Do not be tempted to compare two-out-of-three treatments repeatedly!

Instead, we can apply an analysis of variance (ANOVA), which will allow us to test for the significance of the difference between more than two sample means, on the assumption that the underlying distributions are (approximately) normal (if the underlying distributions are not normal, or unknown, use the non-parametric ANOVA test described in Section 3.4.4).

The null hypothesis again is $\mu_1 = \mu_2 \ldots = \mu_k$. The underlying assumption of the ANOVA test is this: if the k samples are really all taken from the same distribution (null hypothesis), then we should be able to estimate the population variance by two methods: i) by computing the variance among the k sample means ("between-column-variance"), and ii) by computing the variance within each sample individually ("within-column-variance"). If all sample means are indeed from the same population, we should get the same variance, irrespective of how we computed it.

To conduct the analysis of variance, then, we first need to determine the between-column-variance σ_b^2 (Equation 3.9) and the within-column-variance σ_w^2 (Equation 3.10):

$$\sigma_b^2 = \frac{\sum n_j(\mu_j - \overline{\mu})^2}{k - 1} \tag{3.9}$$

n_j is the size of the jth sample, μ_j the mean of the jth sample, $\overline{\mu}$ the grand mean of all samples combined, and k the number of samples (treatments).

$$\sigma_w^2 = \sum \left(\frac{n_j - 1}{n_T - k}\right)\sigma_j^2 \tag{3.10}$$

$n_T = \sum n_j$ is the total sample size, and σ_j^2 is the variance of the jth sample.

Once we have determined these two variances, we determine the F ratio given in Equation 3.11, and test it for significance:

$$F = \frac{\sigma_b^2}{\sigma_w^2} \tag{3.11}$$

Parametric ANOVA: Testing for Significance

To test for significance, we use the F-distribution given in Table 3.18 on page 62. The number of degrees of freedom f_1 of the numerator is given by $f_1 = k - 1$, the number of degrees of freedom f_2 of the denominator is given by $f_2 = n_T - k$. If the F value calculated using Equation 3.11 exceeds the critical value obtained from Table 3.18, the null hypothesis is rejected, meaning that there *is* a significant difference between the k sample means. Otherwise, the null hypothesis that $\mu_1 = \mu_2 \ldots = \mu_k$ is accepted.

Parametric ANOVA Example: Dead End Escape Revisited

In the following example we will revisit the example of the two different robot control programs to achieve dead-end-escape behaviour, given in Section 3.3.4.

For convenience, the times taken to escape from a dead end by either program are given again in Table 3.5.

Table 3.5. Times in seconds taken by two different dead-end-escape programs

Program A	10.2 9.5 9.7 12.1 8.7 10.3 9.7 11.1 11.7 9.1
Program B	9.6 10.1 8.2 7.5 9.3 8.4

In this example, we have $n_A = 10$, $n_B = 6$, $n_T = 16$, $\overline{x_A} = 10.2$, $\sigma_A = 1.1$, $\overline{x_B} = 8.9$, $\sigma_B = 0.98$, $\overline{\mu} = 9.7$ and $k = 2$.

Following Equations 3.9 and 3.10 we get $\sigma_b^2 = \frac{10(10.2-9.7)^2+6(8.9-9.7)^2}{2-1} = 6.3$, $\sigma_w^2 = \frac{10-1}{16-2}1.1^2 + \frac{6-1}{16-2}0.98^2 = 1.1$. This results in an F-value of $F = \frac{6.3}{1.1} = 5.7$.

The degrees of freedom for the statistical analysis are $f_1 = k - 1 = 1$ and $f_2 = n_T - k = 14$. From Table 3.18 we see that the critical value $F_{crit} = 4.6$. The computed F-value exceeds F_{crit}, we therefore reject the null hypothesis, confirming that there is a significant difference between the performance of the two programs.

3.4 Non-Parametric Methods to Compare Samples

Some experiments do not generate actual measurable values (such as speed, time, *etc.*), but merely a ranked performance. For instance, two floor cleaning robots may have been assessed by human volunteers, watching the cleaning operation and assessing the "quality" of performance on some arbitrary scale. The experimenter might want to know whether the performance of both robots is essentially the same, or whether one robot is perceived to be better than the other. Parametric methods such as the t-test cannot be used in this case, because no parametric data is available. It may also be the case that the data to be analysed is not normally distributed, but follows some other distribution. This is where non-parametric methods come in.

Unlike parametric methods, non-parametric methods make no assumptions about the distribution of the data that is being analysed; they are therefore more "robust", because they can be used to analyse data of *any* distribution. They are, however, less powerful (able to reject a null hypothesis) than parametric methods, and if it is established that a parametric method could be used, then that should be the method of choice.

Most non-parametric methods analyse rank, *i.e.* how high or low a score was achieved in a particular task, and compare whether the rank distribution is in agreement with a particular null hypothesis or not. We look at several rank-based non-parametric methods in this section.

3.4.1 Median and Median Confidence Interval

In the case of normally distributed data, we had defined the "expected value", the mean, by Equation 3.2, and given a confidence interval for the mean by Equation 3.4.

For data that is not normally distributed, or for data for which the distribution is unknown, a similar measure can be given, the *median*. The median is simply the central value in our data. If, for instance, the following measurements were obtained from a noisy sensor [18 22 25 29 43 59 67 88 89], then 43 is the central value and therefore the median (for datasets of even length the median is determined as the average of the two central values).

To determine confidence intervals of the median, we can use Table 3.6. This table indicates the lower and upper bound of the confidence interval as the number r of values inwards from the two extreme points of the dataset.

Table 3.6. Confidence interval for median, given as number r of data values inwards from the two extreme values in the data set (5% significance level, after [Barnard et al., 1993])

n	1-5	6 7 8 9 10 11 12 13 14 15 16 17 18 19 20 21 22 23 24 25 26 27 28 29 30
$r_{5\%}$	not avail.	1 1 1 2 2 2 3 3 3 4 4 5 5 5 6 6 6 7 7 8 8 8 9 9 10

In our example ($n=9$) we get $r=2$, the confidence interval at the 5% significance level for the median is therefore [22,88]. As in the case of mean and standard error, the median is often reported with its confidence interval, for example in bar charts, or when cited in the literature.

3.4.2 Non-Parametric Testing for Differences Between Two Distributions (Mann-Whitney U-Test or Wilcoxon Rank Sum Test)

The Mann-Whitney U-statistic (also known as the Wilcoxon rank sum test) determines whether two independent samples[5] have been drawn from the same or two different populations. It can be viewed as the non-parametric counterpart of the t-test for independent samples, and can be used for comparing samples of different or identical sizes.

The null hypothesis of the test is that the median and the shape of both distributions is the same, the alternative hypothesis that it is not. Because the U-test does not require knowledge of any parameters of the underlying distributions, it is particularly useful for small sample sizes (for which it is not possible to determine whether they are normally distributed or not) or for distributions for which it is known that they are not normally distributed. A further strength of the U-statistic is that it can be applied for data that has been measured on some arbitrary scale, as long as the data is ordinal: the U-test is a non-parametric test, making no assumptions about the underlying distributions.

Non-Parametric Testing for Differences: Example 1

As an example, let us assume that a particular household robot is marketed both in Japan and Europe. The manufacturer would like to know whether customer acceptance in these two markets is significantly different (at the 5% significance level), and therefore obtains subjective "satisfaction" scores (0-100 points) from 15 European and 14 Japanese households.

The performance scores obtained are shown in Table 3.7.

Table 3.7. "Customer satisfaction" scores obtained by the robot in the two different markets

Europe 85 60 90 10 33 67 70 76 33 89 95 50 15 45 56
Japan   60 65 30 25 45 70 50 43 35 61 70 30 29 56

The data shown in Table 3.7 is non-Gaussian (not normally distributed), and the Mann-Whitney U test can be used to investigate the null hypothesis that there is no significant difference between the scores obtained in the two markets.

To conduct the test, we first have to rank the performance data, as shown in Table 3.8 (note that for tied ranks the average between the ranks is entered).

---

[5] The Mann-Whitney test can only be used to compare *two* distributions. To compare more than two data sets, the non-parametric ANOVA should be used (Section 3.4.4).

Table 3.8. Ranked scores obtained in Europe and Japan

Rank	Score	Market	Score Europe	Score Japan
1	95	Europe	1	
2	90	Europe	2	
3	89	Europe	3	
4	85	Europe	4	
5	76	Europe	5	
7	70	Europe	7	
7	70	Japan		7
7	70	Japan		7
9	67	Europe	9	
10	65	Japan		10
11	61	Japan		11
12.5	60	Europe	12.5	
12.5	60	Japan		12.5
14.5	56	Europe	14.5	
14.5	56	Japan		14.5
16.5	50	Europe	16.5	
16.5	50	Japan		16.5
18.5	45	Europe	18.5	
18.5	45	Japan		18.5
20	43	Japan		20
21	35	Japan		21
22.5	33	Europe	22.5	
22.5	33	Europe	22.5	
24.5	30	Japan		24.5
24.5	30	Japan		24.5
26	29	Japan		26
27	25	Japan		27
28	15	Europe	28	
29	10	Europe	29	
Total			195	240

We then compute the so-called U-statistic (Equation 3.12) for each sample, the mean of the U-statistic (Equation 3.13) and the standard error of the U-statistic (Equation 3.14), with n_1 and n_2 being the number of items in group 1 and 2 resp., and R_1 and R_2 the sums of ranks for groups 1 and 2 respectively.

$$U_1 = n_1 n_2 + \frac{n_1(n_1 + 1)}{2} - R_1 \tag{3.12}$$

$$U_2 = n_1 n_2 + \frac{n_2(n_2 + 1)}{2} - R_2$$

$$\mu_u = \frac{n_1 n_2}{2} \tag{3.13}$$

$$\sigma_u = \sqrt{\frac{n_1 n_2 (n_1 + n_2 + 1)}{12}} \tag{3.14}$$

In this example, we get the following results:

$$U_1 = 15 \times 14 + \frac{15(15 + 1)}{2} - 195 = 135$$

$$U_2 = 15 \times 14 + \frac{14(14 + 1)}{2} - 240 = 75 \tag{3.15}$$

$$\mu_u = \frac{15 \times 14}{2} = 105$$

$$\sigma_u = \sqrt{\frac{15 \times 14(15 + 14 + 1)}{12}} = 22.9$$

If both n_1 and n_2 are larger than 10, the U-statistic can be approximated by the normal distribution [Levin and Rubin, 1980, p. 486]. This is the case here. For smaller sample sizes, see page 49.

Figure 3.7 shows this situation. We are interested to determine whether the two distributions differ or not at the 5% significance level, *i.e.* whether U is significantly *above or below* the mean. This is therefore a two-tailed test, and we are interested to determine whether U lies inside the acceptance region, or in the 5% of area outside the acceptance region.

Perhaps the easiest way to determine this is to use a table of the standard normal distribution, such as Table 3.2. What we would like to establish is the value z for which the area underneath the normal distribution curve between μ and z equals the size of the acceptance region indicated in Figure 3.7. In our case this area should be 47.5% of the total area underneath the normal distribution curve, because we are establishing significance at the 5% level, and are using a two-tailed test.

We see from Table 3.2 that for $z=1.96$ the area between μ and z underneath the normal distribution curve is 0.475 (47.5% of the area under the curve of the normal distribution), therefore $z=1.96$ is the answer we are looking for in this example.

The lower and upper limits of the acceptance region are then given by Equation 3.16:

$$\mu_u - z\sigma_u < A < \mu_u + z\sigma_u \tag{3.16}$$

where A is the acceptance region, μ_u the mean of the U-statistic given in Equation 3.13 and σ_u the standard deviation of the U-statistic, given in Equation 3.14.

In this case we get $105 - 1.96 \times 22.9 < A < 105 + 1.96 \times 22.9$, which results in an acceptance region of 60.12 to 149.88 (Figure 3.7).

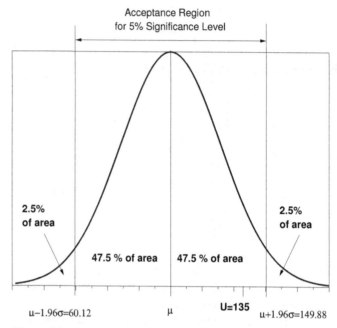

Figure 3.7. Graphical representation of the household-robot-example

U_1=135 and $U_2 = 75$ are clearly inside the acceptance region, and we therefore have no reason to reject the null hypothesis at the 5% level of significance (consequently, we assume that there is no significant difference between the rankings of the robot in the two distributions).

Looking at Table 3.2, we can furthermore see that for 41% of the area underneath the normal distribution curve (*i.e.* a significance level of 18%) we get a z=1.34, resulting in an acceptance region of $105 \pm 1.34 \times 22.9$ (74.3 to 135.7), which means that even at a significance level of 18% we would not reject the null hypothesis that the robot ranks equally highly in terms of customer satisfaction in both markets.

The z value can, of course, also be obtained using readily available software packages. In Scilab one obtains z by typing

```
cdfnor("X",0,1,P,Q)
```

with Q being half of the desired significance level, and P=1-Q. In this particular case this means

```
cdfnor("X",0,1,0.975,0.025)
    ans  =

    1.959964
```

Likewise, the significance level for a given value of U can be determined in Scilab by

```
[Q P]=cdfnor("PQ",U,mu,sigma)
```

with P being half the significance level (two-tailed test!), mu the mean of the U-statistic (Equation 3.13) and sigma the standard error of the U-statistic (Equation 3.14).

U-Test Example 2: Dead End Escape Revisited

In Section 3.3.4 we analysed the dead end escape behaviour achieved, using two different control programs. For convenience, the results obtained both with program A and with program B are shown again in Table 3.9.

Table 3.9. Results obtained using two different dead-end-escape programs

Program A	10.2	9.5	9.7	12.1	8.7	10.3	9.7	11.1	11.7	9.1	Sum
Rank A	5	10	7.5	1	13	4	7.5	3	2	12	65
Program B	9.6	10.1	8.2	7.5	9.3	8.4					
Rank B	9	6	15	16	11	14					71

Here, the size of one sample is smaller than 10, and instead of assuming a normal distribution of the U-statistic, we compare the two values U_1 and U_2 (Equation 3.12) against the critical value U_{crit} for the U-test, given in Table 3.10. If the smaller of $< U_1, U_2 >$ falls below U_{crit}, we reject the null hypothesis.

Table 3.10. Critical values for the U-statistic for sample sizes $n < 10$, at the 5% significance level

	n_1																	
	3	4	5	6	7	8	9	10	11	12	13	14	15	16	17	18	19	20
$n_2 = 3$	-	0	0	1	1	2	2	3	3	4	4	5	5	6	6	7	7	8
$n_2 = 4$	-	0	1	2	3	4	4	5	6	7	8	9	10	11	11	12	13	14
$n_2 = 5$	0	1	2	3	5	6	7	8	9	11	12	13	14	15	17	18	19	20
$n_2 = 6$	1	2	3	5	6	8	10	11	13	14	16	17	19	21	22	24	25	27
$n_2 = 7$	1	3	5	6	8	10	12	14	16	18	20	22	24	26	28	30	32	34
$n_2 = 8$	2	4	6	8	10	13	15	17	19	22	24	26	29	31	34	36	38	41
$n_2 = 9$	2	4	7	10	12	15	17	20	23	26	28	31	34	37	39	42	45	48
$n_2 = 10$	3	5	8	11	14	17	20	23	26	29	33	36	39	42	45	48	52	55

From Equation 3.12 we get

$$U_1 = 10 \times 6 + \frac{10(10+1)}{2} - 65 = 50 \qquad (3.17)$$

$$U_2 = 10 \times 6 + \frac{6(6+1)}{2} - 71 = 10 \qquad (3.18)$$

As $U_2 < U_{crit} = 11$ (Table 3.10), we reject the null hypothesis, confirming our earlier result that there is a significant difference between the performance of the two programs.

3.4.3 The Wilcoxon Test for Paired Observations

This is a variation of the test introduced in Section 3.4.2, which can be used if in experiments where the outcomes of experiments are paired by some criterion.

Paired Observations: Example 1

Let's assume that a robot manufacturer produces two models of domestic robots, and would like to know whether customers in his main market prefer one model over the other. He therefore selects 15 potential customers, and asks them to assess both model 1 (M1) and model 2 (M2) on a subjective scale of 0 to 100, with respect to "customer satisfaction". Scores are "paired" in this case, because each customer evaluates both robots. The results obtained are shown in rows i and ii of Table 3.11.

Table 3.11. "Customer satisfaction" on a subjective scale of 1-100 of robot models M1 and M2

i	M1	54	62	67	42	13	89	56	45	68	23	30	24	35	87	70	
ii	M2	54	38	43	8	22	73	50	48	27	26	13	27	67	66	66	
iii	M1-M2	0	24	24	34	-9	16	6	-3	41	-3	17	-3	-32	21	4	
iv	rank		10.5	10.5	13	6	7	5	2	14	2	8	2	12	9	4	
																	Sum
v	ranks +		10.5	10.5	13		7	5		14		8			9	4	$T_+ = 81$
vi	ranks -					6			2		2		2	12			$T_- = 24$

The null hypothesis is that there is no significant difference between the customers' evaluation of model 1 and model 2, in other words, that the difference between the two satisfaction ratings is not significantly different from zero. The manufacturer would like this hypothesis to be tested at the 10% significance level.

Testing the Hypothesis

To conduct the test, we first need to compute the differences in evaluation for each pair. If the null hypothesis was indeed true, the probability that model 1 obtains a higher score should be the same as the probability that model 2 scores higher. The test is based on this assumption.

Having computed the differences between results (row iii in Table 3.11), we rank the *absolute* values of these differences, in a similar way to the ranking performed in the U-test (Section 3.4.2). Differences of zero are ignored, and the ranks of ties are the average rank over all tied values, as before. Row iv of Table 3.11 shows the results.

We then compute the sum T_+ of all "positive" ranks, and the sum T_- of all "negative" ranks (rows v and vi of Table 3.11).

The expected value μ_T and the standard deviation σ_T are given by Equations 3.19 and 3.20:

$$\mu_T = \frac{n(n+1)}{4} \tag{3.19}$$

$$\sigma_T = \sqrt{\frac{(2n+1)\mu_T}{6}} \tag{3.20}$$

with n being the number of non-zero differences.

In this particular example we get

$$\mu_T = \frac{14(14+1)}{4} = 52.5$$

$$\sigma_T = \sqrt{\frac{(2 \times 14 + 1)52.5}{6}} = 15.9$$

Wilcoxon showed that if the number of non-zero differences n is greater than 8 ($n = 14$ in this case), and the null hypothesis is true, then T_+ and T_- approximately follow a normal distribution [Wilcoxon, 1947]. We can therefore apply a similar technique to that used in Section 3.4.2, and compare T_+ and T_- against the standard normal distribution given in Table 3.1: if T_+ and T_- are outside the acceptance region given by Equation 3.16, the null hypothesis is rejected, otherwise it is accepted.

For the 10% significance level, we find z=1.65 from Table 3.1, which gives an acceptance region for T_+ and T_- of 52.5 ± 1.65 × 15.9 (*i.e.* 26.3 to 78.7). T_+ and T_- are outside the acceptance region, and we therefore reject the null hypothesis at the 10% significance level: customer satisfaction is *not* identical for both robots.

Or, put differently, for $z = 1.79$ we get an acceptance region of 52.5 ± 1.79 × 15.9 (24 to 80.9). The corresponding significance level for $z = 1.79$ for a two-tailed test is 7.4%, meaning that we would reject the null hypothesis at the 7.4% significance level.

In Scilab this significance level can be determined by

```
[Q P]=cdfnor("PQ",TP,mu,sigma)
P   =

    0.0367970
```

with TP being T_+, mu=μ_T, sigma=σ_T and P half the significance level (two-tailed test).

Paired Observations Example 2: Rubber Wheels *vs* Plastic Wheels Revisited

Let us look at the example analysed earlier, the comparison of a robot's speed using rubber tyres *vs* the speed using plastic tyres (Section 3.3.5).

The speeds achieved are shown again in Table 3.12.

Table 3.12. Speeds attained by the same robot, using plastic or rubber wheels respectively, executing the same task for each pair of data

Rubber	58	31	82	87	40	42	71	45	35	23	13	24	
Plastic	59	146	111	111	75	60	57	81	13	59	106	37	
Diff	1	115	29	24	35	18	- 14	36	- 22	36	93	13	Sum
Rank	1	12	7	6	8	4	3 9.5		5 9.5		11	2	
Rank +	1	12	7	6	8	4	9.5		9.5		11	2	T_+=70
Rank -							3		5				T_-=8

Following Equations 3.19 and 3.20 we obtain $\mu_T = \frac{12(12+1)}{4} = 39$ and $\sigma_T = \sqrt{\frac{(2\times 12+1)39}{6}} = 12.7$.

Again using a 5% significance level, we find $z = 1.96$ from Table 3.2, and determine the acceptance interval as $39\pm 1.96\times 12.7 = [14, 64]$. T_+ and T_- are outside the acceptance interval, and, as before, we reject the null hypothesis: the speeds attained with rubber wheels are significantly different to those attained with plastic wheels.

Paired Observations Example 3: Comparing Path Planners

Two versions of a robot path planner (A and B) are compared by running each path planner on the same robot, in the same environment. The results obtained are shown in Table 3.13.

Again, we compute the differences in performance, and rank their absolute values, as shown in Table 3.14. Differences of zero are again ignored, and tied ranks are awarded the average rank.

Table 3.13. Performance indicators of two path planners, being executed on the same robot and performing the same task

A 30 50 40 70 20 50 40 80 70 10 10
B 20 70 30 60 50 50 50 60 20 50 20

Table 3.14. Non-parametric analysis of the data presented in Table 3.13

												Sum
Performance A	30	50	40	70	20	50	40	80	70	10	10	
Performance B	20	70	30	60	50	50	50	60	20	50	20	
Difference	10	-20	10	10	-30	0	-10	20	50	-40	-10	
Rank$_A$	3		3	3		-		6.5	10			$T_A=25.5$
Rank$_B$		6.5			8	-	3			9	3	$T_B=29.5$

Applying Equations 3.19 and 3.20, we get

$$\mu_T = \frac{10(10+1)}{4} = 27.5,$$

$$\sigma_T = \sqrt{\frac{(2 \times 10 + 1)27.5}{6}} = 9.81.$$

In this example, we are interested to determine whether there is a significant difference between the two path planners at the 8% significance level. From Table 3.2 we get $z = 1.75$ for this significance level; the acceptance interval therefore is $27.5 \pm 1.75 \times 9.81$ (*i.e.* 10.3 to 44.7, Equation 3.16). Both T_A and T_B are within the acceptance interval, therefore the null hypothesis (that there is no significant difference between the two path planners) cannot be rejected.

3.4.4 Testing for Difference Between Two and More Groups (Non-Parametric ANOVA, Kruskal Wallis Test)

Section 3.3.6 presented a method of determining whether more than two samples are significantly different or not, assuming that the underlying distributions are normal.

This assumption, however, is not always met, and the non-parametric analysis of variance described in this section can be used in those cases. As in other non-parametric tests, the non-parametric ANOVA is based on rank.

As in the case of parametric ANOVA, discussed in Section 3.3.6, we are interested to determine whether there is a significant difference between k groups of data, or not.

To conduct the test, combine the data from all groups and rank it (tied ranks are awarded the average, as usual). Then add the ranks attained by each group, yielding k values R_k.

We then compute the H statistic given in Equation 3.21:

$$H = \frac{12}{N(N+1)} \left(\sum_{i=1}^{k} \frac{R_i^2}{n_i} \right) - 3(N+1) \tag{3.21}$$

N is the total number of data points in all k groups, and n_i the number of data points in group i.

The sampling distribution of the H statistic is a very close approximation of the χ^2 distribution, if each of the k samples includes at least five observations (even for three observations the relationship is close). Therefore, to determine whether there are any differences at all between the groups, we compare the H-statistic against the critical values given in the χ^2 table (Table 3.24 on page 71), for $k - 1$ degrees of freedom. If H exceeds the critical value from the table, the null hypothesis (that there is no significant difference between groups) is rejected.

3.4.5 Kruskal Wallis Test Example: Dead End Escape Revisited

Earlier we determined whether two different dead-end-escape programs were significantly different, or not. We used the t-test (Section 3.3.4) and the Mann-Whitney U-test (Section 3.4.2). We will now use the Kruskal-Wallis test as a third method to analyse the data given in Table 3.15.

Table 3.15. Results obtained using two different dead-end-escape programs

											Sum
Program A	10.2	9.5	9.7	12.1	8.7	10.3	9.7	11.1	11.7	9.1	
Rank A	5	10	7.5	1	13	4	7.5	3	2	12	65
Program B	9.6	10.1	8.2	7.5	9.3	8.4					
Rank B	9	6	15	16	11	14					71

In this particular example, we have $k = 2$, $N = 16$. H is computed as $H = \frac{12}{16(16+1)} \left(\frac{65^2}{10} + \frac{71^2}{6} \right) - 3(16+1) = 4.7$, following Equation 3.21.

For $k - 1 = 1$ degrees of freedom we get a critical value $\chi_{0.05} = 3.84$ from Table 3.24 on page 71. H exceeds $\chi_{0.05}$, and we therefore reject the null hypothesis and confirm our earlier findings that there is a significant difference between the two dead-end-escape performances.

3.5 Testing for Randomness in a Sequence

It is sometimes interesting to know whether two samples (*e.g.* outcomes of a test, result of some operation, *etc.*) appear in random order, or whether there is order in the way that the two samples are drawn. For example, this information is needed to determine whether a time series is stationary or not (see Section 4.3.2).

By way of example, let us assume we have a floor cleaning robot that is intended to move in a random manner. We log the left and right turns the robot takes, and would like to know whether they are truly randomly distributed, or not. We obtain the following sequence: LLRRLLRRLLRRLLRRLLLRRRLLL-RRRLLLLRRRRLLLLRRRRLLRR. This sequence does not look very random at all, and the following one sample runs test [Bendat and Piersol, 2000] will confirm that indeed this sequence is not random.

In order to conduct the test, we need to determine the numbers n_1 and n_2 of how often symbol 1 (L) and symbol 2 (R) occur, and the number r of runs, *i.e.* sub-sequences of identical symbols. In this case $n_1 = n_2 = 24$, and $r = 18$ (LL-RR-LL-RR-LL-RR-LL-RR-LLL-RRR-LLL-RRR-LLLL-RRRR-LLLL-RRRR-LL-RR).

Mean μ_r and standard error σ_r of the r statistic are given by Equations 3.22 and 3.23. In this case we get $\mu_r = 25$ and $\sigma_r = 3.4$:

$$\mu_r = \frac{2n_1n_2}{n_1 + n_2} + 1 \tag{3.22}$$

$$\sigma_r = \sqrt{\frac{2n_1n_2(2n_1n_2 - n_1 - n_2)}{(n_1 + n_2)^2(n_1 + n_2 - 1)}} \tag{3.23}$$

where n_1 is the number of occurrences of symbol 1, and n_2 the number of occurrences of symbol 2.

The following runs test can be applied if either n_1 or n_2 is larger than 20, because in this case the distribution r of runs follows the normal distribution, which means we can apply our usual procedure of determining a confidence interval within which the number of actually observed runs r would lie if our null hypothesis (H_0: the symbols appear in random order) were correct. We will do the analysis at a significance level of 0.05; Figure 3.8 shows the situation.

The test is intended to determine whether our observed number of runs r is within the acceptance interval around $\mu_r \pm z\sigma_r$, *i.e.* above the lower acceptance limit and below the upper acceptance limit. This is, therefore, a two-tailed test, and the acceptance interval underneath the normal distribution curve covers 95% of the area underneath the curve.

From Table 3.2 we find that for a significance level of 0.05 (*i.e.* for an area between μ_r and the end of the upper confidence interval of 0.475) $z = 1.96$, and the the acceptance interval is therefore $25 \pm 1.96 \times 3.4 = [18.3, 31.7]$. $r = 18$

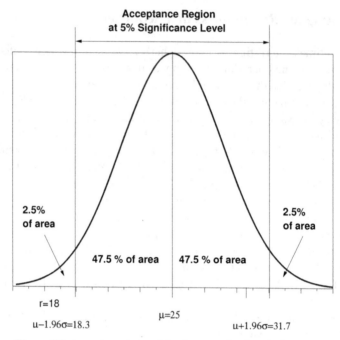

**Acceptance Region
at 5% Significance Level**

2.5%
of area

2.5%
of area

47.5 % of area | 47.5 % of area

r=18

μ=25

u−1.96σ=18.3

u+1.96σ=31.7

Figure 3.8. Acceptance interval for the example given in Section 3.5

is outside that acceptance interval, and we therefore reject the null hypothesis: the turns performed by the robot are not in a random sequence.

The following Scilab code will conduct a runs test for randomness of a sequence consisting of the two symbols "T" and "F":

```
function []=runstest(run)
// Ulrich Nehmzow
// Determines whether a sequence "run" is random or not
// ("run" is a sequence of %T and %F)

// Compute n1,n2 and r
n1=sum(run==%T)
n2=sum(run==%F)
r=1

if((n1<20) & (n2<20))
        printf("There is not enough data to conduct the test\n")
        abort
end
last=run(1)
for i=2:length(run)
        if(last~=run(i))
            r=r+1
            last=run(i)
        end
end
// Now perform the test for randomness
mur=1+(2*n1*n2)/(n1+n2)
sigmar=sqrt((2*n1*n2*(2*n1*n2-n1-n2))/((n1+n2-1)*(n1+n2)^2))

// [P Q]=cdfnor("PQ",r,mur,sigmar) // P/2 is the significance level
```

```
P=input('Please  enter  the  desired  significance  level')
lowerbound=cdfnor("X",mur,sigmar,P/2,1-P/2)
upperbound=cdfnor("X",mur,sigmar,1-P/2,P/2)
printf("The  accept.  interval  for  r  is  %4.1f  to  %4.1f\n",lowerbound,upperbound)
if((r>lowerbound) & (r<upperbound))
        printf("r=%d  is  within  that  region,  therefore  the  number  of  runs
        is  random,  H0  is  accepted.\n",r)
else
        printf("r=%d  is  outside  that  region,  therefore  the  number  of  runs  is  not
        random,  H0  is  rejected.\n",r)
end
```

3.6 Parametric Tests for a Trend (Correlation Analysis)

The previous sections dealt with establishing whether there are statistically significant differences between two (or more) groups, or not. They did not address the question whether two sets of data are correlated, *i.e.* whether knowing the first data set reveals anything about the second data set or not, in other words, whether there is a causal relationship between the two variables or not.

In this section, therefore, we will look at tests that establish exactly that: is one data set predicted by the other data set (and if yes, to what degree), or is there no correlation between the two sets?

In tests for a trend one selects one of the two data sets, X, as the independent variable, and establishes to what degree the other data set, Y, is dependent on X. Y is, not surprisingly, called the dependent variable.

3.6.1 Parametric Linear Regression Analysis

The tests discussed so far did not investigate whether there was a causal relationship between two sets of data. However, in many experimental situations such a relationship exists. For instance, people's weight and height are linked in a causal relationship (they are "correlated"). If one of the two variables is chosen as the independent variable (*i.e.* selected by the user), the other variable is dependent upon that choice: if I identify the height of a person as 1.80 m, then his weight will be "determined" by that within some confidence interval.

Regression analysis investigates the relationship between independent and dependent variable. It is a parametric test, and assumes that the dependent variable is normally distributed (this is not necessary for the independent variable).

For example, let us assume we have measured speed and battery charge of a mobile robot, and obtained the values given in Table 3.16. Plotted against each other, these values result in the scatter-plot shown in Figure 3.9.

To the naked eye it seems obvious that there is a causal relationship between battery charge and robot speed, indicated by the line drawn through the data. We would like to know what this relationship is mathematically, and whether it is significant or not.

Table 3.16. Example: battery charge in relation to robot speed

Charge [Volts]	12.0	11.7	11.5	11.1	10.8	10.4
Speed [cm/s]	41	45	41	40	32	34

Charge [Volt]

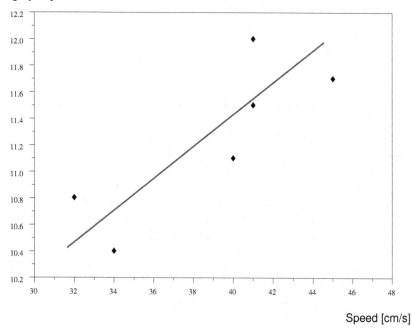

Speed [cm/s]

Figure 3.9. Example: relationship between battery charge and speed of a mobile robot

Linear Regression

Given the independent variable (battery charge), we can predict the dependent variable (robot speed), using the linear relationship indicated in Figure 3.9. Linear regression is the method of determining parameters a and b in the equation for a straight line given in Equation 3.24:

$$Y = aX + b \qquad (3.24)$$

with Y being the dependent variable, X the independent variable, and a and b slope and intercept point defined in Equations 3.25 and 3.26:

$$b = \frac{\sum XY - n\overline{XY}}{\sum X^2 - n\overline{X}^2} \qquad (3.25)$$

$$a = \overline{Y} - b\overline{X} \qquad (3.26)$$

with X and Y again the independent and dependent variable resp., $\overline{X}$ and $\overline{Y}$ their means, and n the number of pairs of independent and dependent variables available.

In this example the independent variable (battery charge) was measured as given in Table 3.16. Following Equations 3.25 and 3.26 this results in a linear relationship between charge and speed as given in Equation 3.27:

$$speed = 6.7 \times charge - 36.9 \tag{3.27}$$

Instead of computing a and b by hand, using Equations 3.25 and 3.26, they can of course also be computed using mathematical packages. In Scilab the following does the trick:

```
-->[a b]=reglin(charge,speed)
 b   =

    - 36.906103
 a   =

    6.7323944
```

This is a prediction of robot speed, given the battery charge, and the next question that arises is: how good is this prediction?

The *standard error of estimate* s_e, defined in Equation 3.28, does this. It measures the scatter of the observed values around the regression line:

$$s_e = \sqrt{\frac{\sum(Y - \hat{Y})^2}{n-2}} = \sqrt{\frac{\sum Y^2 - a\sum Y - b\sum XY}{n-2}} \tag{3.28}$$

with X being the values of the independent variable, Y being a value of the dependent variable, $\hat{Y}$ being a value of the dependent variable, estimated using Equation 3.24, and n the number of pairs of independent and dependent variable used to obtain the regression line.

The larger the standard error of estimate s_e, the larger the scatter around the regression line, and the weaker the correlation between independent and dependent variable. For $s_e = 0$, on the other hand, we expect a perfect prediction of the dependent variable, given the independent variable.

In the specific example given here, we get a standard error of estimate of $s_e = \sqrt{\frac{39.15}{6-2}} = 3.13$, as shown in Table 3.17.

Similar to the standard deviation defined earlier, and assuming that the observed points are normally distributed around the regression line, we will expect to find 68% of all points in the interval of $\pm 1 s_e$ around the regression line, 95.5%

Table 3.17. Computing the standard error of estimate in the example given in Table 3.16

X	Y	$\hat{Y} = 6.7X - 36.9$	$(Y - \hat{Y})^2$
12	41	43.5	6.25
11.7	45	41.5	12.32
11.5	41	40.2	0.72
11.1	40	37.5	6.40
10.8	32	35.5	11.97
10.4	34	32.8	1.49
			$\sum 39.15$

in an interval of $\pm 2s_e$ around the regression line, and 99.7% of all points in the interval $\pm 3s_e$ around the regression line.

This allows us to make statements concerning the confidence we have in the prediction made. If, for instance, we would like to predict the speed of the robot, given a battery charge of 11.3 Volt, we predict (Equation 3.27) speed=6.7 × charge -36.9=38.8. We can now say that we are 68% certain the robot's speed will be in the interval 38.8 ± 3.13, or 95.5% certain that it will be in the interval $38.8 \pm 2 \times 3.13$.

Strictly speaking, these calculations are only applicable to sample sizes of $n > 30$, because for smaller sample sizes it is incorrect to apply the prediction intervals of the normal distribution, and the conclusions drawn here are therefore inaccurate. However, they demonstrate the mechanism. To correct for smaller sample sizes, one needs to take the t-distribution shown in Table 3.3. If, for instance, we would like to be 95% certain that the true speed of the robot lies within the computed confidence interval, we find t=2.776 for $k = n - 2 = 4$ degrees of freedom and 5% significance level in Table 3.3. With 95% certainty, therefore, the robot's speed will be within the interval $38.8 \pm 2.776 \times 3.13$.

One final note on relationships between independent and dependent variable that are not linear. Although linear regression analysis assumes a linear relationship, and tests for significance based on that assumption, it *can* be used for nonlinear relationships, too, by transforming the data sets so that a linear relationship is established. For instance, log-transforming will linearise an exponential relationship between independent and dependent variable.

Linear Regression: Testing for Significance (F-Statistic)

There are two tests for significance that can be applied to linear regression: i) to test whether variance in the dependent variable is accounted for by corresponding changes in variance of the independent variable (F-test) and ii) to test whether the difference between regression line and data points (the "error") differs significantly from zero (t-test). Both of these tests will be discussed in the following — if either fails, there is no significant causal relationship between independent and dependent variable, and the regression line is best not used as a predictor.

F-Statistic

The F-statistic allows us to test for significance and tests whether variance in the dependent variable Y is accounted for by variance in the independent variable X. F is determined by Equation 3.29:

$$F = \frac{RSS}{DMS} \qquad (3.29)$$

$$RSS = \frac{S_{XY}^2}{S_{XX}} \qquad (3.30)$$

$$DMS = \frac{DSS}{n-2} \qquad (3.31)$$

$$DSS = S_{YY} - \frac{S_{XY}^2}{S_{XX}} \qquad (3.32)$$

$$S_{XX} = \sum X^2 - \frac{(\sum X)^2}{n} \qquad (3.33)$$

$$S_{YY} = \sum Y^2 - \frac{(\sum Y)^2}{n} \qquad (3.34)$$

$$S_{XY} = \sum XY - \frac{(\sum X)(\sum Y)}{n} \qquad (3.35)$$

with n the number of data pairs used in the regression analysis.

The resulting F value is checked against the critical values given in Table 3.18, the degrees of freedom are $f_1 = 1$ for the numerator, and $f_2 = n - 2$ for the denominator. Table 3.18 gives the critical values for the 5% significance level. To be significant, computed F values must be greater than those shown in Table 3.18. Values for other significance levels can be found in statistical tables, or computed. In Scilab, this is done by the following command:

```
F=cdff("F",DFN,DFD,Q,P)
```

with DFN being the degrees of freedom of the numerator, DFD the degrees of freedom of the denominator, P the significance level and Q=1-P.

Coming back to the example given in Table 3.16, we get $S_{XX} = 1.775$, $S_{YY} = 118.83$, and $S_{XY} = 11.95$. From Equation 3.29 follows $F = 8.38$. The critical value of F for $f_1 = 1$ and $f_2 = 6 - 2 = 4$ is $F_{crit} = 7.71$ (Table 3.18). The computed F value of 8.38 is greater than that, indicating that the correlation between battery charge and robot speed is significant at the 5% level.

t-Test

We can also test whether a predicted and an actually observed value differ significantly or not, using the t-statistic defined in Equation 3.36:

Table 3.18. Critical values for the F distribution (p=0.05). f_1 values are the degrees of freedom of the numerator, f_2 those of the denominator

f_2 \ f_1	1	2	3	4	5	6	7	8	9	10	11	12	13	14	15	16	17	18	19	20	∞
1	161.45	18.51	10.13	7.71	6.61	5.99	5.59	5.32	5.12	4.96	4.84	4.75	4.67	4.60	4.54	4.49	4.45	4.41	4.38	4.35	3.84
2	199.50	19.00	9.55	6.94	5.79	5.14	4.74	4.46	4.26	4.10	3.98	3.89	3.81	3.74	3.68	3.63	3.59	3.55	3.52	3.49	3.00
3	215.71	19.16	9.28	6.59	5.41	4.76	4.35	4.07	3.86	3.71	3.59	3.49	3.41	3.34	3.29	3.24	3.20	3.16	3.13	3.10	2.60
4	224.58	19.25	9.12	6.39	5.19	4.53	4.12	3.84	3.63	3.48	3.36	3.26	3.18	3.11	3.06	3.01	2.96	2.93	2.90	2.87	2.37
5	230.16	19.30	9.01	6.26	5.05	4.39	3.97	3.69	3.48	3.33	3.20	3.11	3.03	2.96	2.90	2.85	2.81	2.77	2.74	2.71	2.21
6	233.99	19.33	8.94	6.16	4.95	4.28	3.87	3.58	3.37	3.22	3.09	3.00	2.92	2.85	2.79	2.74	2.70	2.66	2.63	2.60	2.10
7	236.77	19.35	8.89	6.09	4.88	4.21	3.79	3.50	3.29	3.14	3.01	2.91	2.83	2.76	2.71	2.66	2.61	2.58	2.54	2.51	2.01
8	238.88	19.37	8.85	6.04	4.82	4.15	3.73	3.44	3.23	3.07	2.95	2.85	2.77	2.70	2.64	2.59	2.55	2.51	2.48	2.45	1.94
9	240.54	19.38	8.81	6.00	4.77	4.10	3.68	3.39	3.18	3.02	2.90	2.80	2.71	2.65	2.59	2.54	2.49	2.46	2.42	2.39	1.88
10	241.88	19.40	8.79	5.96	4.74	4.06	3.64	3.35	3.14	2.98	2.85	2.75	2.67	2.60	2.54	2.49	2.45	2.41	2.38	2.35	1.83
11	242.98	19.40	8.76	5.94	4.70	4.03	3.60	3.31	3.10	2.94	2.82	2.72	2.63	2.57	2.51	2.46	2.41	2.37	2.34	2.31	1.79
12	243.91	19.41	8.74	5.91	4.68	4.00	3.57	3.28	3.07	2.91	2.79	2.69	2.60	2.53	2.48	2.42	2.38	2.34	2.31	2.28	1.75
13	244.69	19.42	8.73	5.89	4.66	3.98	3.55	3.26	3.05	2.89	2.76	2.66	2.58	2.51	2.45	2.40	2.35	2.31	2.28	2.25	1.72
14	245.36	19.42	8.71	5.87	4.64	3.96	3.53	3.24	3.03	2.86	2.74	2.64	2.55	2.48	2.42	2.37	2.33	2.29	2.26	2.22	1.69
15	245.95	19.43	8.70	5.86	4.62	3.94	3.51	3.22	3.01	2.85	2.72	2.62	2.53	2.46	2.40	2.35	2.31	2.27	2.23	2.20	1.67
16	246.46	19.43	8.69	5.84	4.60	3.92	3.49	3.20	2.99	2.83	2.70	2.60	2.51	2.44	2.38	2.33	2.29	2.25	2.21	2.18	1.64
17	246.92	19.44	8.68	5.83	4.59	3.91	3.48	3.19	2.97	2.81	2.69	2.58	2.50	2.43	2.37	2.32	2.27	2.23	2.20	2.17	1.62
18	247.32	19.44	8.67	5.82	4.58	3.90	3.47	3.17	2.96	2.80	2.67	2.57	2.48	2.41	2.35	2.30	2.26	2.22	2.18	2.15	1.60
19	247.69	19.44	8.67	5.81	4.57	3.88	3.46	3.16	2.95	2.79	2.66	2.56	2.47	2.40	2.34	2.29	2.24	2.20	2.17	2.14	1.59
20	248.01	19.45	8.66	5.80	4.56	3.87	3.44	3.15	2.94	2.77	2.65	2.54	2.46	2.39	2.33	2.28	2.23	2.19	2.16	2.12	1.57
21	248.31	19.45	8.65	5.79	4.55	3.86	3.43	3.14	2.93	2.76	2.64	2.53	2.45	2.38	2.32	2.26	2.22	2.18	2.14	2.11	1.56
22	248.58	19.45	8.65	5.79	4.54	3.86	3.43	3.13	2.92	2.75	2.63	2.52	2.44	2.37	2.31	2.25	2.21	2.17	2.13	2.10	1.54
23	248.83	19.45	8.64	5.78	4.53	3.85	3.42	3.12	2.91	2.75	2.62	2.51	2.43	2.36	2.30	2.24	2.20	2.16	2.12	2.09	1.53
24	249.05	19.45	8.64	5.77	4.53	3.84	3.41	3.12	2.90	2.74	2.61	2.51	2.42	2.35	2.29	2.24	2.19	2.15	2.11	2.08	1.52
25	249.26	19.46	8.63	5.77	4.52	3.83	3.40	3.11	2.89	2.73	2.60	2.50	2.41	2.34	2.28	2.23	2.18	2.14	2.11	2.07	1.51
26	249.45	19.46	8.63	5.76	4.52	3.83	3.40	3.10	2.89	2.72	2.59	2.49	2.41	2.33	2.27	2.22	2.17	2.13	2.10	2.07	1.50
27	249.63	19.46	8.63	5.76	4.51	3.82	3.39	3.10	2.88	2.72	2.59	2.48	2.40	2.33	2.27	2.21	2.17	2.13	2.09	2.06	1.49
28	249.80	19.46	8.62	5.75	4.50	3.82	3.39	3.09	2.87	2.71	2.58	2.48	2.39	2.32	2.26	2.21	2.16	2.12	2.08	2.05	1.48
29	249.95	19.46	8.62	5.75	4.50	3.81	3.38	3.08	2.87	2.70	2.58	2.47	2.39	2.31	2.25	2.20	2.15	2.11	2.08	2.05	1.47
30	250.10	19.46	8.62	5.75	4.50	3.81	3.38	3.08	2.86	2.70	2.57	2.47	2.38	2.31	2.25	2.19	2.15	2.11	2.07	2.04	1.46
∞	254.3	19.50	8.53	5.63	4.36	3.67	3.23	2.93	2.71	2.54	2.40	2.30	2.21	2.13	2.07	2.01	1.96	1.92	1.88	1.84	1.00

$$t = \left| \frac{Y - \hat{Y}}{s_e} \right| \qquad (3.36)$$

with Y being an observed measurement of the dependent variable, $\hat{Y}$ a predicted value, and s_e the standard error of the estimate defined in Equation 3.28. Let's assume we actually measured the robot's speed as 36.2, having a charge of 11.3 Volt. Is this difference between actually measured value (36.2) and predicted value (38.8) significant or not at the 5% level? Following Equation 3.36, we get a t-value of $t = \left| \frac{36.2 - 38.8}{3.13} \right| = 0.8$. From Table 3.3 we get a critical value ($n - 2 = 6 - 2 = 4$ degrees of freedom) of 2.776. However, 0.8 is below this critical value, and therefore there is no significant difference between the prediction and the actually measured value.

3.6.2 Pearson's Linear Correlation Coefficient r

The linear correlation coefficient r (Pearson's r), which measures the association between two continuous variables x and y, is given by Equation 3.37:

$$r = \frac{\sum_i (x_i - \bar{x})(y_i - \bar{y})}{\sqrt{\sum_i (x_i - \bar{x})^2} \sqrt{\sum_i (y_i - \bar{y})^2}} \qquad (3.37)$$

with r being Pearson's linear correlation coefficient, and $\bar{x}$ and $\bar{y}$ the means of the x_is and the y_is. r assumes a value between -1 (perfect negative correlation —

points lie on a perfectly straight line with negative slope) to +1 (perfect positive correlation — points lie on a perfectly straight line with positive slope), with small values of r indicating that there is no strong linear correlation between x and y.

Testing for Significance of Pearson's r

Even two random variables might have a non-zero r, and the question we would like to answer is whether a specific r signifies a statistically *significant* correlation or not.

Table 3.19 gives the significance levels for Pearson's correlation coefficient r for $df = N - 2$ degrees of freedom (N is the number of data pairs). If we are interested whether r is significantly above or below zero, we have to use a two-tailed test, otherwise a one-tailed test.

Table 3.19. Significance levels for Pearson's r

df	Level of significance for one-tailed test			
	.05	.025	.01	.005
	Level of significance for two-tailed test			
	.10	.05	.02	.01
1	.988	.997	.9995	.9999
2	.900	.950	.980	.990
3	.805	.878	.934	.959
4	.729	.811	.882	.917
5	.669	.754	.833	.874
6	.622	.707	.789	.834
7	.582	.666	.750	.798
8	.549	.632	.716	.765
9	.521	.602	.685	.735
10	.497	.576	.658	.708
15	.412	.482	.558	.606
20	.360	.423	.492	.537
25	.323	.381	.445	.487
30	.296	.349	.409	.449
40	.257	.304	.358	.393
60	.211	.250	.295	.325
80	.183	.217	.256	.284
100	.164	.195	.230	.254

Linear Correlation Coefficient: Scilab Code

The following Scilab program will compute Pearson's linear correlation coefficient.

```
function [r]=pearson(a,b)
// Ulrich Nehmzow
// Computes the linear correlation coefficient (Pearson's r)
// between two time series a and b

a=a-mean(a)
b=b-mean(b)
numerator=0
for i=1:length(a)
```

```
              numerator=numerator+a(i)*b(i)
end
r=numerator/(sqrt(sum(a^2))*sqrt(sum(b^2)))

// Now analyse for statistical significance
// Note: this program approximates the function for the
// critical values for degrees of freedom beyond df=60,
// if the result is "close", the result should be checked with
// tabulated critical values

rcrit=[0.997000 0.950000 0.878000 0.811000 0.754000 0.707000
0.666000 0.632000 0.602000 0.576000 0.553000 0.532000 0.514000
0.497000 0.482000 0.468000 0.456000 0.444000 0.433000 0.423000
0.413000 0.404000 0.396000 0.388000 0.381000 0.374000 0.367000
0.361000 0.355000 0.349000 0.344000 0.339000 0.335000 0.330000
0.325000 0.321000 0.317000 0.312000 0.308000 0.304000 0.301000
0.298000 0.294000 0.291000 0.288000 0.285000 0.282000 0.279000
0.276000 0.273000 0.271000 0.268000 0.266000 0.264000 0.262000
0.259000 0.257000 0.255000 0.252000 0.250000]

df=length(a)-2
if df<=60
        critical=rcrit(df)
else
        critical=0.195+exp(-0.05*df)
end
printf("The critical value for statistical
            significance is  %5.2f\n",critical)
if abs(r)>critical
        printf("Therefore the computed r of
            %5.2f is significant (p<5\%, two-tailed)\n",r)
else
        printf("Therefore the computed r of
            %5.2f is not significant (p>5\%, two-tailed)\n",r)
end
```

Linear Correlation Coefficient (Pearson's r): Example

For the example mentioned in Section 3.6.1 on page 57, where we were interested to see whether there was a correlation between robot speed and battery charge, the Pearson correlation coefficient is computed as $r = 0.823$. We would like to know whether this r is significant at the 5% significance level.

Because we want to know whether r is either significantly smaller or bigger than zero, a two-tailed test applies. For $df = N - 2 = 4$ we find a critical value of 0.811. The r in this case is just above that critical value, and is therefore significant.

3.7 Non-Parametric Tests for a Trend

3.7.1 Spearman Rank Correlation

Section 3.6.2 presented correlation analysis as a measure of closeness of association between two variables, based on numerical values, and assuming an underlying normal distribution.

Some experiments, however, do not generate numerical data for analysis, and correlation analysis cannot be applied in those cases. But provided *rankings* regarding size or merit are available, a correlation analysis is still possible: the Spearman rank correlation analysis discussed in this section.

Besides being applicable to situations where only rankings are available, this rank correlation analysis has the further advantages that it is computationally much cheaper than correlation analysis based on numerical values, is much less influenced by extreme values in the data than analysis based on numerical values, and, in certain cases, can be used to detect non-linear correlation as well.However, rank correlation should not be used if it is possible to calculate the linear correlation coefficient given in Equation 3.37. The Spearman rank correlation test computes a rank correlation coefficient r_s between -1 (perfect correlation, negative slope) and +1 (perfect correlation, positive slope), with the null hypothesis being that there is no correlation between the data sets ($r_s = 0$). If a significant correlation is detected, this means that the rank correlation coefficient r_s differs significantly from zero, and that one of the two data sets can be used to make a meaningful prediction of the other.

When there are no ties in the data, r_s is given by Equation 3.38[6]:

$$r_s = 1 - \frac{6 \sum d^2}{n(n^2 - 1)} \tag{3.38}$$

with d being the difference between the ranks for each pair of observations, and n the total number of paired observations. Equation 3.38 can be used whether the distributions of the two data sets are normal or not.

Spearman Rank Correlation Example: Cleaning Robot

A floor cleaning robot has four different types of behaviour: moving in a spiral, moving in a straight line, avoiding obstacles and following a wall. The data given in Table 3.20 shows how often wall following behaviour was observed in a particular experiment, and how often the right bump sensor of the robot was triggered. Figure 3.10 shows the scatter diagram for this data. The question is: is this relationship significantly different from $r_s = 0$ (the null hypothesis) or not?

To compute r_d, we rank both columns of Table 3.20 individually, and compute the difference of ranks for each pair. This is shown in Table 3.21. As always

---

[6] Tied ranks introduce an error to Equation 3.38.

Table 3.20. Experiment with floor cleaning robot: relationship between signals on the robot's right bumper and wall following behaviour

Experiment Nr.	Nr of right bumps	Nr of wall follows
1	18	4
2	12	3
3	21	6
4	13	3
5	22	7
6	40	9
7	38	12
8	8	4
9	41	12

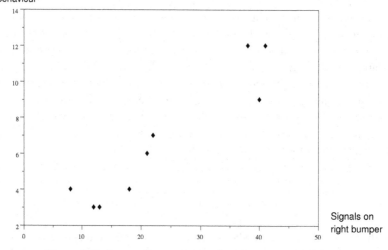

Figure 3.10. Scatter diagram for the data shown in Table 3.20

in rank analyses, tied ranks are awarded the average rank (*i.e.* 3rd and 4th rank tied each get rank "3.5", the next rank after that is "5").

With $n=9$ we get $r_s = 1 - \frac{6*12.5}{9(9^2-1)} = 0.896$. Is this a significant deviation from $r_s = 0$, our null hypothesis?

Spearman Rank Correlation: Testing for Significance

To determine significance in the Spearman rank correlation test, one applies one of two criteria: for $n < 30$ one computes the acceptance region using Table 3.22. For larger n the sampling distribution of r_s is approximately normal, and we can use Table 3.2.

Table 3.21. Experiment with floor cleaning robot: relationship between signals on the robot's right bumper and wall following behaviour (raw data, ranks and rank differences)

Experiment Nr.	Nr of right bumps	Nr of wall follows	Rank bumps	Rank WF	d^2
1	18	4	4	3.5	0.25
2	12	3	2	1.5	0.25
3	21	6	5	5	0
4	13	3	3	1.5	2.25
5	22	7	6	6	0
6	40	9	8	7	1
7	38	12	7	8.5	2.25
8	8	4	1	3.5	6.25
9	41	12	9	8.5	0.25
					$\sum$ 12.5

We will first look at the case $n < 30$. This is the case of the floor cleaning robot; we will therefore analyse the data given above.

Our null hypothesis is that there is *no* statistically significant correlation in the ranked data, the alternative hypothesis is that there *is* a significant correlation in the ranked data. We investigate these hypotheses at the 5% significance level. Because we would consider both too small and too large r_s as beyond expectation, a two-tailed test is appropriate.

Table 3.22 shows that for a $n = 9$ pairs the critical values for r_s are ± 0.68, *i.e.* the lower limit of the acceptance region is -0.68, the upper limit is 0.68. Our computed rank correlation coefficient of $r_s = 0.896$ is outside the acceptance region, we therefore reject the null hypothesis that $r_S = 0$, meaning that there is indeed a statistically significant correlation between bumps on the right bumper sensor and the robot executing wall following behaviour.

Table 3.22. Critical values for Spearman's rank correlation r_s for two-tailed tests (5% significance level) (after [Levin and Rubin, 1980]). n is the number of data pairs

n	1	2	3	4	5	6	7	8	9	10	11	12	13	14	15
					.90	.83	.75	.71	.68	.64	.61	.58	.55	.53	.52

n	16	17	18	19	20	21	22	23	24	25	26	27	28	29	30
	.50	.49	.47	.46	.45	.44	.42	.42	.41	.40	.39	.38	.37	.37	.36

If n is larger than 30, the sampling distribution of r_s is approximately normal, and we can use the standard error σ_{r_s} given in Equation 3.39 to compute acceptance intervals for r_s.

$$\sigma_{r_s} = \frac{1}{\sqrt{n-1}}$$ (3.39)

By way of example, let's assume that we continued the experimentation with the floor cleaning robot for another 31 experiments, so that $n=40$. r_s evaluates to 0.53.

For a two-tailed test — we are interested whether the rank correlation coefficient is significantly below *or above* the null hypothesis of $r_s = 0$ — at a significance level of 5%, we find from Table 3.2 (page 33) $z=1.96$. The limits of our acceptance region are $0 \pm z\sigma_{r_s}$. In this particular case, the lower limit of our acceptance region is $-1.96\frac{1}{\sqrt{40-1}} = -0.31$, the upper limit of the acceptance region is 0.31. The computed r_s of 0.53 is outside that acceptance region, and again the null hypothesis of $r_s = 0$ is rejected.

Spearman Rank Correlation: Scilab Code

The following program computes the Spearman rank correlation coefficient and tests for significance:

```
function[rs]=spearmanr(x,y)
getf('ranking.sci')
// Ulrich Nehmzow
// Computes the non-parametric Spearman rank correlation coefficient
// between vectors x and y

rankx=ranking(x)
ranky=ranking(y)
d=rankx-ranky
n=length(x)
rs=1-((6*sum(d^2))/(n^3-n))
printf("Spearman rank correlation coefficient: %f\n",rs)

// Now performing analysis for significance
if n<4
        printf("Too little data for significance analysis\n")
        break
elseif n<31
        critical=[ 999 999 999 999 .90 .83 .75 .71 .68 .64 .61 .58 .55
.53 .52 .50 .49 .47 .46 .45 .44 .42 .42 .41 .40 .39 .38 .37 .37 .36]
        if(abs(rs)>critical(n))
                printf("This coefficient is significant (p<5\%)\n")
        else
                printf("This coefficient is not significant (p>5\%)\n")
        end
else
        sigma=1/sqrt(n-1)
        if abs(rs)>1.96*sigma
                printf("This coefficient is significant (p<5\%)\n")
        else
                printf("This coefficient is not significant (p>5\%)\n")
        end
end
```

```
function [rnk]=ranking(a)
// Ulrich Nehmzow
// Performs a ranking of vector a
// as preprocessing for non-parametric statistical tests

[s k]=sort(a)

// Now produce the ranking vector
for pos=1:length(a)
```

```
//          r(k(pos))=pos
            r(pos)=pos
end

// Now check for tied ranks and adjust the r-vector accordingly
pos=1
while pos <= length(a)-1
        ties=1
        while s(pos)==s(pos+ties)
                ties=ties+1
                if pos+ties>=length(a)
                        break
                end
        end
r(pos:pos+ties-1)=mean(r(pos:pos+ties-1))
pos=pos+ties
end

// Now reallocate ranks to position in original input vector
for pos=1:length(a)
        rnk(k(pos))=r(pos)
end
```

3.8 Analysing Categorical Data

Mean, standard deviation, t-test and many other statistical analysis methods can only be applied to continuous-valued data. In robotics experiments, however, there are many experiments in which results are obtained as "categories", for example in classification systems, whose task it is to allocate sensory data to one of several categories. In this section, we will look at methods of analysing such categorical data.

Contingency Tables

Nominal variables are defined as variables that are members of an unordered set, such as for example "colour" or "taste". It is nominal variables that we consider here.

For the following considerations, we are interested in determining whether two nominal variables are associated or not. This question is relevant for example for classification tasks, where one variable is the input signal, and one the output. In this case, the question asked is "is the output of the classifier associated with the input signals?", in other words, "is the classifier doing a good job?".

Data of two variables can be displayed in a contingency table, which will allow us to perform a so-called crosstabulation analysis. For example, if there was a robot contest, in which three robots compete a number of times in three different disciplines, a contingency table which would state how often each robot won each contest could be built, and crosstabulation analysis could be used to determine whether there was a correlation between robot and discipline. This would establish whether any robot was particularly good at any specific discipline. Figure 3.23 shows the contingency table for this analysis.

Table 3.23. Example of a contingency table. $n_{A,X}$ is the number of times robot X won contest A, $N_{.A}$ the total number of winners in contest A, $N_{Z.}$ the total number of wins of robot Z, *etc.*

	Contest A	Contest B	Contest C	
Robot X	$n_{A,X}$	$n_{B,X}$	...	
Robot Y			...	
Robot Z	...	...	...	$N_{Z.}$
	$N_{.A}$	$N_{.B}$	$N_{.C}$	N

3.8.1 Determining the Association Between Two Variables (χ^2 Test)

One test to determine the significance of an association between two variables is the χ^2 test.

Let N_{ij} be the number of events where the variable x has value i and variable y has value j. Let N be the total number of events. Let $N_{i.}$ be the number of events where x has value i, regardless of y, and $N_{.j}$ the number of events where y has value j, regardless of the value of x:

$$N_{i.} = \sum_j N_{ij} ,$$
$$N_{.j} = \sum_i N_{ij} ,$$
$$N = \sum_i N_{i.} = \sum_j N_{.j} .$$

Deriving the Table of Expected Values

The null hypothesis in the χ^2 test is that the two variables x and y have no significant correlation. In order to test this null hypothesis, "expected values" need to be determined, to express what values we expect to obtain if the null hypothesis were true. The expected values can either be derived from general considerations dependent on the application, or from the following reasoning.

In a table such as Table 3.23, $\frac{n_{ij}}{N_{.j}}$ is an estimate of the probability that a certain event i happens, given j, *i.e.* $\frac{n_{ij}}{N_{.j}} = p(i|j)$. If the null hypothesis were true, the probability for a particular value of i, given a particular value of j should be exactly the same as the probability of that value of i regardless of j, *i.e.* $\frac{n_{ij}}{N_{.j}} = p(i|j) = p(i)$.

It is also true that $p(i) = \frac{N_{i.}}{N}$. Under the assumption that the null hypothesis is true we can therefore conclude that

$$\frac{n_{ij}}{N_{.j}} = \frac{N_{i.}}{N} \tag{3.40}$$

which yields the table of expected values n_{ij}:

$$n_{ij} = \frac{N_{i.}N_{.j}}{N} \tag{3.41}$$

χ^2 is defined in Equation 3.42:

$$\chi^2 = \sum_{i,j} \frac{(N_{ij} - n_{ij})^2}{n_{ij}} \tag{3.42}$$

The computed value for χ^2 (see Equation 3.42) in conjunction with the $\chi^2_{.05}$ probability function (Table 3.24) can now be used to determine whether the association between variables i and j is significant or not. For a table of size I by J, the number of degrees of freedom m is

$$m = IJ - I - J + 1 \tag{3.43}$$

If $\chi^2 > \chi^2_{.05}$ (see Table 3.24) there is a significant correlation between the variables i and j.

Table 3.24. Table of critical χ^2 values, for significance levels of 0.1, 0.05 and 0.01

DOF	10%	5%	1%
1	2.71	3.84	6.63
2	4.61	5.99	9.21
3	6.25	7.81	11.34
4	7.78	9.49	13.28
5	9.24	11.07	15.09
6	10.64	12.59	16.81
7	12.02	14.07	18.48
8	13.36	15.51	20.09
9	14.68	16.92	21.67
10	15.99	18.31	23.21
11	17.28	19.68	24.72
12	18.55	21.03	26.22
13	19.81	22.36	27.69
14	21.06	23.68	29.14
15	22.31	25.00	30.58
16	23.54	26.30	32.00
17	24.77	27.59	33.41
18	25.99	28.87	34.81
19	27.20	30.14	36.19
20	28.41	31.41	37.57

If m is greater than 30, significance can be tested by calculating $\sqrt{2\chi^2} - \sqrt{2m - 1}$. If this value exceeds 1.65, there is a significant correlation between i and j.

Instead of using tables such as Table 3.24, the critical values for the χ^2 distribution can also be computed. In Scilab this is done by

```
cdfchi("X",DF,Q,P)
```

with DF being the number of degrees of freedom, P the significance level chosen, and Q=1-P.

Practical Considerations Regarding the χ^2 Statistic

In order for the χ^2 statistic to be valid, the data needs to be well conditioned. Two rules of thumb determine when this is the case:

1. In the n_{ij} table of expected values, no cell should have values below 1. In cases where $m \geq 8$ and $N \geq 40$ no values must be below 4 ([Sachs, 1982, p. 321]).
2. In the n_{ij} table of expected values, not more than 5% of all values should be below 5.

If either of the above conditions is violated, rows or columns of the contingency table can be combined to meet the two criteria given above.

Example χ^2 Test: Assessing Self-Localisation

A mobile robot is placed in an environment that contains four prominent landmarks, A, B, C and D. The robot's landmark identification program produces four responses, α, β, γ and δ to the sensory stimuli received at these four locations. In an experiment totalling 200 visits to the various landmarks, contingency Table 3.25 is obtained (numbers indicate the frequency of a particular map response obtained at a particular location).

Table 3.25. Contingency table obtained for landmark-identification program

	α	β	γ	δ	
A	19	10	8	3	$N_{A.} = 40$
B	7	40	9	4	$N_{B.} = 60$
C	8	20	23	19	$N_{C.} = 70$
D	0	8	12	10	$N_{D.} = 30$
	$N_{.\alpha} = 34$	$N_{.\beta} = 78$	$N_{.\gamma} = 52$	$N_{.\delta} = 36$	N=200

Is the output of the classifier significantly associated with the location the robot is at?

Answer: Following Equation 3.41 $n_{A\alpha} = \frac{40*34}{200} = 6.8$, $n_{A\beta} = \frac{40*78}{200} = 15.6$, and so on (the table of expected values is Table 3.26).

The table of expected values is well conditioned for the χ^2 analysis; no values are below 4.

Following Equation 3.42, $\chi^2 = \frac{(19-6.8)^2}{6.8} + \frac{(10-15.6)^2}{15.6} + \ldots = 66.9$. The system has $16-4-4+1 = 9$ degrees of freedom (Equation 3.43). $\chi^2_{0.05} = 16.9$, according to Table 3.24. The inequality

Table 3.26. Table of expected values

	α	β	γ	δ
A	6.8	15.6	10.4	7.2
B	10.2	23.4	15.6	10.8
C	11.9	27.3	18.2	12.6
D	5.1	11.7	7.8	5.4

$$\chi^2 = 66.9 > \chi^2_{0.05} = 16.9 \tag{3.44}$$

holds; therefore there *is* a significant association between robot location and output of the location identification system.

3.8.2 Determining the Strength of an Association: Cramer's V

The χ^2 test is a very general test in statistics, and as such has limited expressive power. In fact, provided the number of samples contained in a contingency table is large enough, the test will often indicate a significant correlation between the variables. This has to do with the "power" of the test, which will amplify even small correlations beyond the "significance" level, provided enough samples are available. For this reason, it is better to re-parametrise χ^2 so that it becomes independent from the sample size. This will allow us to assess the strength of an association, and to compare contingency tables with one another.

Cramer's V re-parametrises χ^2 to the interval $0 \leq V \leq 1$. $V = 0$ means that there exists no association between x and y, $V = 1$ means perfect association. V is given by Equation 3.45:

$$V = \sqrt{\frac{\chi^2}{N min(I-1, J-1)}} \tag{3.45}$$

with N being the total number of samples in the contingency table of size $I \times J$, and $min(I - 1, J - 1)$ being the minimum of $I - 1$ and $J - 1$.

The following Scilab code will compute χ^2 and Cramer's V, given a contingency table m:

```
function [X2]=chisq (m)
// Ulrich Nehmzow
// Performs a Chi squared analysis for the contingency table m

[r c]=size (m)
N=sum (m)

// Compute the table of expected values
for i=1:r
```

```
        for  j=1:c
                n(i,j)=sum(m(:,j))*sum(m(i,:))/N
        end
end

// Now compute contributions to Chi squared
for  i=1:r
        for  j=1:c
                xcontrib(i,j)=((m(i,j)-n(i,j))^2)/n(i,j)
        end
end
X2=sum(xcontrib)
[p sig]=cdfchi("PQ",X2,r*c-r-c+1)
if  (sig>0.05)
        printf("There  is  no  significant  correlation
        between  the  two  variables  (p>0.05)\n")
else
        printf("There  is  a  significant  correlation
        between  the  two  variables  (p<%4.3f)\n",sig)
end
v=sqrt(X2/(sum(m)*(-1+min(size(m)))))
printf("Cramer''s  V=%f\n",v)
```

Example: Cramer's V

Two different map-building paradigms are to be compared. Paradigm A yields a contingency table as given in Table 3.27, paradigm B produces the table shown in Table 3.28.

Table 3.27. Results of map-building mechanism 1

	α	β	γ	δ	
A	29	13	5	7	$N_{A.} = 54$
B	18	4	27	3	$N_{B.} = 52$
C	8	32	6	10	$N_{C.} = 56$
D	2	7	18	25	$N_{D.} = 52$
	$N_{.\alpha} = 57$	$N_{.\beta} = 56$	$N_{.\gamma} = 56$	$N_{.\delta} = 45$	N=214

The question is: which of the two mechanisms produces a map with a stronger correlation between robot location and map response?

We use Cramer's V to answer that question.

The tables of expected values are given in Tables 3.29 and 3.30. Looking at both tables of expected values, one can see that the data is well conditioned and meets the criteria listed on page 72.

Table 3.28. Results of map-building mechanism 2

	α	β	γ	δ	ϵ	
A	40	18	20	5	7	$N_{A.} = 90$
B	11	20	35	10	3	$N_{B.} = 79$
C	5	16	10	39	5	$N_{C.} = 75$
D	2	42	16	18	9	$N_{D.} = 87$
E	6	11	21	9	38	$N_{D.} = 85$
	$N_{.\alpha} = 64$	$N_{.\beta} = 107$	$N_{.\gamma} = 102$	$N_{.\delta} = 81$	$N_{.\epsilon} = 62$	N=416

Table 3.29. Expected values for map-building mechanism 1

	α	β	γ	δ
A	14.4	14.1	14.1	11.4
B	13.9	13.6	13.6	10.9
C	14.9	14.7	14.7	11.8
D	13.9	13.6	13.6	10.9

In the case of map-building mechanism 1, we determine $\chi^2 = 111$ and $V = 0.42$, in the case of mechanism 2 we obtain $\chi^2 = 229$ and $V = 0.37$. Map 1 has the stronger correlation between map response and location. Both experiments are subject to some random variation, however, so that it is necessary to run each experiment a number of times, to eliminate the influence of random noise.

Table 3.30. Expected values for map-building mechanism 2

	α	β	γ	δ	ϵ
A	13.8	23.1	22.1	17.5	13.4
B	12.2	20.3	19.4	15.4	11.8
C	11.5	19.3	18.4	14.6	11.2
D	13.4	22.4	21.3	16.9	13
E	13.1	21.9	20.8	16.6	12.7

3.8.3 Determining the Strength of Association Using Entropy-Based Measures

The χ^2 analysis and Cramer's V allow us to determine whether or not there is a significant association between rows and columns of a contingency table.

However, what we would also like is some measure of the *strength* of the association. Two quantitative measures of the strength of an association will therefore be discussed below.

The particular scenario we have in mind here is this: a mobile robot explores its environment, constructs a map, and uses this map subsequently for localisation.

Whenever the robot is at some physical location L, therefore, its localisation system will generate a particular response R, indicating the robot's assumed position in the world. In a perfect localisation system, the association between L and R will be very strong, in a localisation system based on random guesswork the strength of the association between L and R will be non-existent, zero.

Entropy based measures, in particular the entropy H and the uncertainty coefficient U, can be used to measure the strength of this association. They are defined as follows.

Using Entropy to Determine the Strength of Association Between Nominal Variables

Table 3.31. Example Contingency Table. The rows correspond to the response produced by the particular localisation system under investigation, and the columns to the "true" location of the robot as measured by an observer. This table represents 100 data points, and also shows the totals for each row and column

	\multicolumn{5}{c}{Location (L)}	
R	0	2	15	0	1	18
e	10	10	0	0	0	20
s	0	2	1	0	19	22
p	5	7	3	1	1	17
.	0	0	0	23	0	23
(R)	15	21	19	24	21	100

In the example given in Table 3.31, a sample consisting of 100 data points has been collected. Each data point has two attributes; one corresponding to the location predicted by the robot (the robot's *response*, R), and the other to the actual location of the robot measured by an observer (the robot's *true location*, L). For example, Table 3.31 shows one cell containing 19 data points where the robot's response was measured as row 3 and the location as column 5.

For contingency table analysis, first of all the row totals $N_{r.}$ for each response r, column totals $N_{.l}$ for each location l and the table total N are calculated according to Equations 3.46, 3.47 and 3.48 respectively. N_{rl} is the number of data points contained in the cell at row r and column l:

$$N_{r.} = \sum_{l} N_{rl} \qquad (3.46)$$

$$N_{.l} = \sum_r N_{rl} \tag{3.47}$$

$$N = \sum_{r,l} N_{rl} \tag{3.48}$$

The row probability $p_{r.}$, column probability $p_{.l}$ and cell probability p_{rl} can then be calculated according to Equations 3.49, 3.50 and 3.51:

$$p_{r.} = \frac{N_{r.}}{N} \tag{3.49}$$

$$p_{.l} = \frac{N_{.l}}{N} \tag{3.50}$$

$$p_{rl} = \frac{N_{rl}}{N} \tag{3.51}$$

The entropy of L, $H(L)$, the entropy of R, $H(R)$ and the mutual entropy of L and R, $H(L, R)$ are given by Equations 3.52, 3.53 and 3.54 respectively:

$$H(L) = -\sum_l p_{.l} \ln p_{.l} \tag{3.52}$$

$$H(R) = -\sum_r p_{r.} \ln p_{r.} \tag{3.53}$$

$$H(L, R) = -\sum_{r,l} p_{rl} \ln p_{rl} \tag{3.54}$$

When applying Equations 3.52, 3.53 and 3.54, bear in mind that $\lim_{p \to 0} p \ln p = 0$.

For the scenario described above, the most important question we would like to have an answer for is this: "Given a particular response R of the robot's localisation system, how certain can we be about the robot's current location L?" This is the entropy of L given R, $H(L \mid R)$. If, on the other hand, one particular location elicits different responses $R1$ and $R2$ on different visits, we don't care. The important point for robot self-localisation is that each response R is strongly associated with exactly one location L.

$H(L \mid R)$ is obtained as follows:

$$H(L \mid R) = H(L, R) - H(R) \tag{3.55}$$

where

$$0 \le H(L \mid R) \le H(L) \tag{3.56}$$

This last property (Equation 3.56) means that the range of values for $H(L \mid R)$ will be dependent on the size of the environment, because $H(L)$ increases as the number of location bins increases.

Using the Uncertainty Coefficient to Determine the Strength of Association Between Nominal Variables

The entropy H is a number between 0 and $ln\ N$, where N is the number of data points. If H is 0, the association between L and R is perfect, *i.e.* each response R indicates exactly one location L in the world. The larger H becomes, the weaker is the association between L and R.

The uncertainty coefficient U provides yet another way of expressing the strength between row and column variables in a contingency table, and it has two very attractive properties: first of all, U always lies between 0 and 1, irrespective of the size of the contingency table. This allows comparisons between tables of different size. Second, the uncertainty coefficient is 0 for a nonexistent association, and 1 for a perfect association. This is intuitively the "right" way round (the stronger the association, the larger the number).

The uncertainty coefficient U of L given R, $U(L\mid R)$, is given as

$$U(L\mid R) \equiv \frac{H(L) - H(L\mid R)}{H(L)} \tag{3.57}$$

A value of $U(L\mid R) = 0$ means that R provides no useful information about L, and implies that the robot's response never predicts its true location. A value of
$U(L\mid R) = 1$ means that R provides all the information required about L, and implies that the response always predicts the true location. It should also be noted that the ordering of the rows and columns in the contingency table makes no difference to the outcome of this calculation.

For the symmetric, general case, the uncertainty coefficient $U(x, y)$ is given by Equation 3.58:

$$U(x, y) = 2\frac{H(y) + H(x) - H(x, y)}{H(x) + H(Y)} \tag{3.58}$$

Example: Computing the Uncertainty Coefficient

A robot localisation system produces the responses shown in Table 3.31. Is there a statistically significant correlation between the system's response, and the robot's location?

In order to answer this question, we compute the uncertainty coefficient $U(L\mid R)$, according to Equation 3.57. To do this, we need to compute $H(L)$, $H(R)$ and $H(L\mid R)$.

By applying Equations 3.52, 3.53, 3.54 and 3.57 we obtain

$$H(L) = -(\tfrac{15}{100}ln\tfrac{15}{100} + \tfrac{21}{100}ln\tfrac{21}{100} + \ldots + \tfrac{21}{100}ln\tfrac{21}{100}) = 1.598$$

$$H(R) = -(\tfrac{18}{100}ln\tfrac{18}{100} + \tfrac{20}{100}ln\tfrac{20}{100} + \ldots + \tfrac{23}{100}ln\tfrac{23}{100}) = 1.603$$

$$H(L,R) = -(0 + \tfrac{2}{100}ln\tfrac{2}{100} + \tfrac{15}{100}ln\tfrac{15}{100} + \ldots + \tfrac{23}{100}ln\tfrac{23}{100} + 0) = 2.180$$

$$H(L \mid R) = 2.180 - 1.603 = 0.577$$

$$U(L \mid R) = \tfrac{1.598-0.577}{1.598} = 0.639.$$

This is an uncertainty coefficient that indicates a fairly strong correlation between the the robot's location and the localisation system's response.

The following Scilab program computes the uncertainty coefficient:

```
function [u]=uc(m)
// Ulrich Nehmzow
// Computes the uncertainty coefficient U(x,y) for the
// contingency table m of size [x y]
[x y]=size(m)
n=sum(m)

// Compute the necessary probabilities
for i=1:x
        pi(i)=sum(m(i,:))
end
pi=pi/n

for j=1:y
        pj(j)=sum(m(:,j))
end
pj=pj/n
pij=m/n

// Now compute the entropies
hx=0
for i=1:x
        if (pi(i)~=0)
                hx=hx-pi(i)*log(pi(i))
        end
end
hy=0
for j=1:y
        if(pj(j)~=0)
                hy=hy-pj(j)*log(pj(j))
        end
end
hxy=0
for i=1:x
        for j=1:y
                if(pij(i,j)~=0)
                        hxy=hxy-pij(i,j)*log(pij(i,j))
                end
```

```
            end
end
// Now compute the uncertainty coefficient
u=2*((hy+hx−hxy)/(hx+hy))
```

3.9 Principal Component Analysis

This chapter has so far discussed how data can be analysed and compared, using statistical means. In some cases, however, one is not as much interested in
whether two data sets differ or not, but rather in visualising a single data set in order to understand more about the data's structure and properties. In these cases,
it can often help to re-present the data in such a way that underlying patterns
become more visible, for example through an appropriate coordinate transformation. Principal component analysis (PCA) is one such method.

Principal component analysis allows the identification of patterns in data, and
can also be used very successfully for data compression. It is in essence a method
of coordinate transformation, representing the original data in such a way that
the standard deviation of the data is largest along the newly defined coordinates.
These new coordinates are the principal components.

Consider the data distribution shown in Figure 3.11 and Table 3.32.

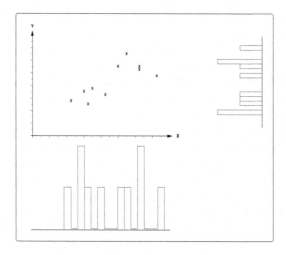

Figure 3.11. Example data for principal component analysis. Histograms show the distribution
along the x and y axes

Table 3.32. Numerical values of the data shown in Figure 3.11

x 0.9 1.2 1.7 1.3 1.4 2.2 2.5 2. 2.9 2.5
y 1.1 1.4 1.3 1. 1.5 2.6 2.2 2.2 1.9 2.1

There are clearly two clusters visible to the naked eye in the data, but using the coordinates (x, y), as shown in Figure 3.11, neither x nor y can be used very well to identify which cluster a data point belongs to. This is indicated by the x and y histogram shown in Figure 3.11. It is, however, possible, to use a coordinate transform to make the two clusters very easy to distinguish, this is shown in Figure 3.12.

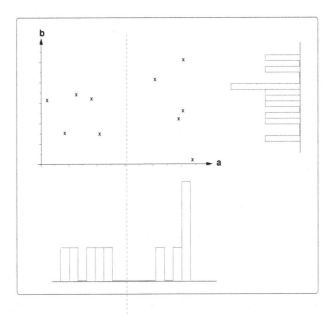

Figure 3.12. The same data as shown in Figure 3.11, but presented along different coordinates (the principal components). The a axis is the first principal component, the b axis the second. The two clusters are now clearly distinguishable by using the a coordinate alone

To perform the PCA, data sets have to have zero mean; we therefore first subtract the mean of x and the mean of y from x and y respectively.

The principal components — the coordinates into which we wish to transform our data — are the eigenvectors of the covariance matrix of the data; we therefore need to determine the covariance matrix, and then compute eigenvectors and their eigenvalues.

The covariance matrix C is defined as shown in Equation 3.59, with the covariance cov given in Equation 3.60:

$$C = \begin{pmatrix} cov(x,x) & cov(x,y) & cov(x,z) & \dots \\ cov(y,x) & cov(y,y) & cov(y,z) & \dots \\ cov(z,x) & cov(z,y) & cov(z,z) & \dots \\ \dots & \dots & \dots & \dots \end{pmatrix} \tag{3.59}$$

$$cov(x,y) = \frac{\sum_{i=1}^{n}(x_i - \bar{x})(y_i - \bar{y})}{n-1} \tag{3.60}$$

with $\bar{x}$ and $\bar{y}$ being the mean of x and y resp. and n the number of data points in x and y. The covariance $cov(x,y)$ indicates whether x and y increase (positive covariance) or decrease (negative covariance) together, or whether the two dimensions are independent from each other (zero covariance). The covariance matrix contains all $\frac{n!}{2(n-2)!}$ covariances that can be computed from n-dimensional data.

Eigenvalue λ_j and eigenvector $\mathbf{w}_j$ of a matrix $\mathbf{A}$ are defined by Equation 3.61:

$$(\mathbf{A} - \lambda_j \mathbf{I_n})\mathbf{w}_j = 0 \tag{3.61}$$

with $\mathbf{I_n}$ being the identity matrix, which has the value "1" along the diagonal, and zero everywhere else.

To compute eigenvalues and eigenvectors manually, using Equation 3.61 is difficult for matrices larger than 3x3, and best performed using mathematical packages.

In `Scilab`, the covariance matrix `covarmat` can be constructed as follows:

```
covarmat=[1*corr(xa,xa,1)/(1-1)  1*corr(xa,ya,1)/(1-1);
1*corr(ya,xa,1)/(1-1)  1*corr(ya,ya,1)/(1-1)]
```

with `xa` and `ya` being the zero-mean original data, and `1` the number of data points in `xa` and `ya`.

The eigenvalues can be determined by the command `eigenvalues=spec(covarmat)`, and the eigenvectors, the principal components `pcas`, by using `[eigenvalue,pcas,bs]=bdiag(covarmat)`.

The following program computes the eigenvectors and their eigenvalues for two-dimensional data, and performs a coordinate transform of the original data into PCA space.

```
function[pcone,pctwo]=pcc(x,y)
// (c) Ulrich Nehmzow
// Computes the principal components of 2D data x and y
// Then converts the original data from Cartesian space into
// the space of the two principal components
```

```
xbasc()
subplot(211)
plot2d(x,y,-2)
xtitle('Original Data','X','Y')

// First convert data into zero mean data
xa=x-mean(x)
ya=y-mean(y)

// Now compute covariance matrix
l=length(xa)
covarmat=[l*corr(xa,xa,1)/(l-1) l*corr(xa,ya,1)/(l-1);
l*corr(ya,xa,1)/(l-1) l*corr(ya,ya,1)/(l-1)]

// Compute the Eigenvalues
eigenvalues=spec(covarmat)

// Compute the Eigenvectors, which are the PCAs
[eigenvalue,pcas,bs]=bdiag(covarmat)
printf("One principal component is ")
printf("%f%f(EV=%f)\n\n\n",pcas(1,1),pcas(2,1),eigenvalue(1,1))
printf("The other principal component is ")
printf("\n%f %f (EV=%f)",pcas(1,2),pcas(2,2),eigenvalue(2,2))

p1=pcas(:,1)
p2=pcas(:,2)
if eigenvalue(1,1) < eigenvalue(2,2)
        temp=p1
        p1=p2
        p2=temp
end

// Now recompute the coordinates
for i=1:length(x)
        pcone(i)=[x(i);y(i)]'*p1
        pctwo(i)=[x(i);y(i)]'*p2
end

subplot(212)
plot2d(pcone,pctwo,-2)
xtitle('Data after coordinate transform','PC 1','PC2')
```

Applying this program to the data presented in Table 3.32 yields the following result:

```
pcc(x,y);
One principal component is
0.791798 0.610783 (EV=0.650954)
```

```
The other principal component is
-0.610783   0.791798  (EV=0.080713)
```

As said above, principal component analysis can be used to identify patterns in data (for example by visualising data along the first principal component), and (lossy) data compression. In the example given in Figures 3.11 and 3.12, compression can be achieved by representing data by using only one coordinate — the first principal component — rather than the original two coordinates.

Further Reading

- Richard Levin and David Rubin, *Applied elementary statistics*, Prentice Hall 1980
- Allen Edwards, *Statistical methods*, Holt, Rinehart and Winston 1967
- Chris Barnard, Francis Gilbert and Peter McGregor, *Asking questions in biology*, Longman 1993

4

Dynamical Systems Theory and Agent Behaviour

Summary. This chapter introduces concepts of dynamical systems theory such as phase space, phase space reconstruction and analysis of phase space, and their application to the analysis of behaving agents.

4.1 Introduction

(Physical) systems whose behavioural descriptors are time-dependent are referred to as *dynamical systems*. Their behaviour can be described through statistical measures and mathematical expressions (in particular, differential or difference equations): the system's behaviour is described mathematically as motion through "state space" (or "phase space"), the multi-dimensional space of the system's descriptors. The behaviour of a mobile robot is clearly a function of time, *i.e.* mobile robots are dynamical systems. It is therefore interesting to apply dynamical systems theory to the analysis of robot-environment interaction. This is the purpose of this chapter.

4.2 Dynamical Systems Theory

Dynamical systems theory is the mathematical theory describing dynamical systems, attempting to describe the behaviour of complex dynamical systems (such as the interaction of an agent with its environment) through differential equations. This behaviour is, broadly speaking, an agent's motion through space, where "space" in the first instance means "physical space", but could also mean other spaces, such as the concept of phase space that will be introduced in Section 4.2.1. The concepts of dynamical systems theory have even been expanded to described agents' motion through "cognitive space", in an attempt to describe cognitive reasoning mathematically.

As a discipline, dynamical systems theory encompasses methods to visualise an agent's motion through space, to analyse and classify it, to measure it and to describe it mathematically. It is therefore an ideal tool to achieve the objectives outlined in Chapter 2, namely to analyse and describe the behaviour of an agent (such as a mobile robot) mathematically.

4.2.1 Phase Space

"Phase space" is the term used to describe that space that describes all possible states of a dynamical system. A particular state in phase space describes the system fully, it contains all information about the system needed to make a prediction of future states of the dynamical system under investigation.

The Phase Space of The Ideal Pendulum

An ideal pendulum, for instance, has one degree of freedom — the arc ϕ along which it is swinging, and the knowledge of the pendulum's position $\phi(t)$ and its velocity $\dot{\phi}(t)$ describes the motion of the pendulum fully, for all times t. The phase space of the ideal pendulum, therefore, is the two-dimensional space defined by $\phi(t)$ and $\dot{\phi}(t)$, and the physical motion of the pendulum can be fully described by the motion through that phase space.

It turns out that the phase space of the ideal pendulum is an ellipse (see Figure 4.1). As the pendulum swings backward and forward in physical space, its $(\phi, \dot{\phi})$ coordinates in phase space move from $(\phi_{max}, 0)$ through $(0, -\dot{\phi}_{max})$, $(-\phi_{max}, 0)$ and $(0, \dot{\phi}_{max})$ back to $(\phi_{max}, 0)$.

The trajectory ("orbit") through phase space — in the pendulum's case the ellipse shown in Figure 4.1 — is referred to as the "attractor", because the dynamical system will follow that particular orbit, irrespective of initial conditions — it is "attracted" to that orbit through phase space.

The Phase Space of Hamiltonian Systems

The phase space of a dynamical system following Hamiltonian mechanics is defined as position $z(t)$ and impulse $\mathbf{p(t)} = m\mathbf{v}(t)$ ($\mathbf{v}$ is the velocity) along each degree of freedom the system has. This is a strict definition of "phase space" and "degree of freedom"; however, the term "degree of freedom" is sometimes also simply used to mean "a single coordinate of phase space". Unless otherwise stated, this is the notion used in this book.

It follows that if a Hamiltonian system has n degrees of freedom, its phase space has $2n$ dimensions, position and impulse along each of the n degrees of freedom of the system.

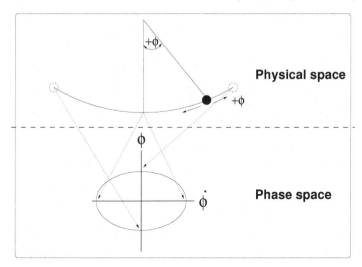

Figure 4.1. Physical movement and phase space of an ideal pendulum. *Arrows* indicate how physical space and phase space relate to each other

Degrees of Freedom of a Mobile Robot

It is not always straightforward to state how many degrees of freedom a system has, and therefore what the size of its phase space is. Fortunately, this knowledge is not needed to reconstruct the attractor (see Section 4.2.3).

In the case of mobile robots, moving in a two-dimensional plane, however, one *can* say something about the degrees of freedom available to the robot. In the following, we will examine four different fundamental types of mobile robot: a fully holonomic robot (which we will refer to as "ball", because its motions are equivalent to those of a ball), a robot with differential drive, a robot with "Ackermann steering" (which we will refer to as "car", because its motions are equivalent to those of a conventional car), and a tracked robot (referred to as "train"). These four types of robot are shown in Figure 4.2.

The first three of these robots are all capable of assuming *any* position and orientation $< x, y, \phi >$ in space, but the means by which they can do so are different in the three cases.

We define a full degree of freedom as an axis (x, y or ϕ) along which any position can be assumed without altering the positions along the remaining two axes. An equivalent definition would be that a full degree of freedom is an axis along which an external force can be applied without meeting resistance (assuming an "ideal" robot).

Following that definition, one can see that in the case of the ball, forces can be applied along all three axes independently, and the position along each of the three axes can be altered without altering the position in the other two axes. The ball therefore has three degrees of freedom.

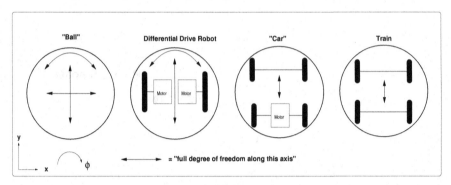

Figure 4.2. Different types of robot drive systems and their full degrees of freedom

The differential drive robot can change its position freely along the y and ϕ axes, but to change position along the x axis, the robot also has to change position along the ϕ axis at the same time. There are therefore two full degrees of freedom (y and ϕ), but because the robot is able to change position along the x axis as well, it has more than two (but less than three) degrees of freedom. Commonly this is referred to as "2.5 degrees of freedom".

In the case of the car, it can change its position freely along the y axis (one degree of freedom). But to change position along the ϕ axis, movement along the y axis is needed, so this isn't a full of degree of freedom. And in order to change position along the x axis, movement along the ϕ axis (and therefore also along the y axis) is needed, so this isn't a full degree of freedom either. How many degrees of freedom does a car have? Perhaps one and two-halves?

Finally, the train has only one full degree of freedom, y, and is incapable of assuming arbitrary positions in x and ϕ — positions along these two axes are pre-determined by the track. It therefore has one degree of freedom, and its phase space is two-dimensional (y and $\dot{y}$).

4.2.2 Illustration: Analysis of Robot Behaviour
Through Phase Space Reconstruction

The following hypothetical example is intended to illustrate the methods employed later in this book, to serve as a illustration of what we are trying to achieve by applying dynamical systems theory to mobile robotics.

Assume that a mobile robot is moving in some environment, perhaps along a trajectory similar to that shown in Figure 4.8. The three variables that describe the robot's trajectory fully are position $x(t)$ and $y(t)$ and heading $\phi(t)$. As in all dynamical systems, these three variables can be described through differential equations. Furthermore, in the robot's case the three variables are coupled, because the robot's control program, the physics of motion and the influence of the environment will mean that x, y and ϕ cannot change completely independently from each other.

In a real robot, we do not know which differential equations describe the robot's motion, but let us assume, for argument's sake, a particular robot's motion was defined by the set of Equations 4.1:

$$\frac{dx}{dt} = \dot{x} = -(y + \phi) \qquad (4.1)$$

$$\frac{dy}{dt} = \dot{y} = x + 0.15y$$

$$\frac{d\phi}{dt} = \dot{\phi} = 0.2 + xz - 10z$$

The differential Equations 4.1 can be solved, for example numerically using the Runge-Kutta method, which will result in the functions $x(t)$, $y(t)$ and $\phi(t)$ shown in Figure 4.3.

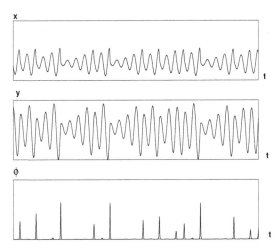

Figure 4.3. The solution of the differential Equations 4.1

It is interesting to plot x, y and ϕ against each other (Figure 4.4). This is the phase space of the system defined by Equation 4.1, *i.e.* that space that defines the system under investigation fully, and allows prediction of the system's future states.

As we said earlier, in a real mobile robot it is usually not possible to determine equations like those given in Equation 4.1. However, it *is* possible to reconstruct the robot's phase space through a method called time-lag embedding (discussed below in Section 4.2.3)! Figure 4.5 shows the reconstructed attractor, and reveals a close similarity to the "real" attractor shown in Figure 4.4.

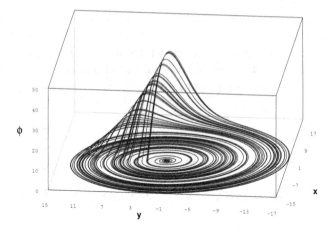

Figure 4.4. The phase space of the system defined by Equation 4.1

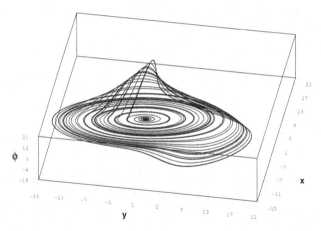

Figure 4.5. The reconstructed phase space of the system defined by Equation 4.1

In summary, the method of analysing the dynamics of robot-environment interaction presented in this chapter is as follows:

1. Observe the robot's motion in $< x, y, \phi >$ space over time.
2. Reconstruct the phase space of that motion, using the method described in Section 4.2.3.
3. Analyse the reconstructed attractor for its dynamical properties, such as sensitivity to initial conditions or dimension of the attractor.

4.2.3 Reconstruction of the Attractor

Our first step towards analysis of a dynamical system, then, is to reconstruct its phase space. Using theoretical considerations, this is not always as easy as it was in the case of the ideal pendulum. Fortunately, however, phase space can be reconstructed from a time series $x(t)$ of observations of the physical system, through a method called time-lag embedding [Peitgen et al., 1992, Kantz and Schreiber, 1997, Abarbanel, 1996]. Figure 4.5 shows the result of this method when applied to the dynamical system defined by Equation 4.1.

Time-Lag Embedding: Illustration

Suppose we had a dynamical system defined by the two differential equations given in Equation 4.2 (see [Kaplan and Glass, 1995] for a full discussion of this example):

$$\dot{x} = y \qquad\qquad (4.2)$$
$$\dot{y} = -bx$$

This is system is fully described by the variables $x(t)$ and $y(t)$, they form the phase space of the system.

In keeping with our "observation of robot behaviour" scenario, let us assume that we observe the behaviour of this system by logging a time series $S(t) = x(t)$ at discrete points t in time. To reconstruct the phase space of this system from $S(t)$, we obviously need to reconstruct only $y(t)$, because $x(t)$ is already given through $S(t)$.

Equation 4.2 indicates that $y = \frac{dx}{dt} = \frac{dS}{dt}$. Therefore, the phase space can be reconstructed by plotting $\dot{S}(t)$ *vs* $S(t)$.

The derivative of a variable $x(t)$ is given by Equation 4.3.

$$\frac{dx}{dt} = \lim_{h \to 0} \frac{x(t+h) - x(t)}{h} \qquad\qquad (4.3)$$

Following the definition given in Equation 4.3, we can approximate $\dot{S}(t)$ by Equation 4.4:

$$\frac{dS(t)}{dt} = \dot{S}(t) = \frac{S(t+h) - S(t)}{h} \qquad\qquad (4.4)$$

Because we have logged S at discrete times t, estimating $\dot{S}$ means applying Equation 4.4, using a suitable h, and consequently reconstructing the phase space means plotting $S(t)$ *vs* $\frac{S(t+h) - S(t)}{h}$. Because all the information needed is contained in $S(t)$ and $S(t+h)$, it is sufficient to simply plot $S(t)$ *vs* $S(t+h)$!

This method can be extended to higher dimensions than two, and therefore be used to reconstruct higher dimensional attractors. This will be discussed next.

Reconstruction of the Attractor Through Time-Lag Embedding

Again, suppose we measure some descriptive element of the agent's behaviour over time, for example the movement of the agent in $< x, y >$ space, obtaining two time series $x(t)$ and $y(t)$. The attractor $\mathbf{D(t_n)}$ — the trajectory taken through phase space — can then be reconstructed through time-lag embedding as given in Equation 4.5:

$$\mathbf{D(t_n)} = (x(t_n - (p-1)\tau), x(t_n - (p-2)\tau), \ldots x(t_n - \tau), x(t_n)) \quad (4.5)$$

with $x(t)$ being a sequential set of measurements (the time series), p being the embedding dimension and τ being the embedding lag.

In order to reconstruct the system's phase space through time lag embedding from an observed time series, therefore, two parameters need to be chosen: the embedding dimension p and the embedding lag τ.

Choosing the embedding dimension. There are three possible scenarios: (i) the embedding dimension chosen is too small to reconstruct the attractor, (ii) it is "just right", or (iii) it is too large. Only the first case will result in errors, because an attractor whose dimension is larger than the chosen embedding dimension cannot be fully unfolded, which means that points that are distant in time end up as close neighbours in phase space (because these neighbours in space are distant in time they are referred to as "false nearest neighbours"). If the embedding dimension is the same or just slightly larger than the dimension of the attractor, reconstruction is obviously no problem. If the embedding dimension chosen is much larger than the attractor's dimension, there is theoretically no problem — the attractor can be reconstructed perfectly — but there are practical (computational and accuracy) reasons why this case is undesirable. It is therefore preferable to select the minimum embedding dimension.

An established method to determine a suitable embedding dimension is to use the false nearest neighbours method discussed in [Kennel et al., 1992]. This method determines the number of false nearest neighbours (close in the reconstructed phase space, but far apart in time) in the reconstructed phase space — when this number is near zero, the attractor is properly unfolded and contains no self-intersections.

Choosing the embedding lag. The second variable to be chosen for the time lag embedding method is the embedding lag τ. The right choice of τ means determining that point at which the sample $x(t + \tau)$ of the observed time series contains new information, compared with $x(t)$. For example, if a slow moving system is sampled at a high sampling frequency, τ is going to be large, because it will take many samples before $x(t + \tau)$ actually contains new information. On the other hand, if the sampling rate is low with respect to the motion of the system, τ is going to be small.

First of all, there is a qualitative method to see the influence of increasing τ. For a small τ, $x(t)$ and $x(t + \tau)$ are essentially identical. If they are plotted against each other, therefore, all points would lie on the diagonal identity line. As τ increases, the reconstructed attractor will expand away from the identity line. This expansion gives us an indication about a suitable choice of τ [Rosenstein et al., 1994].

There are two further ways to determine the point in time at which $x(t)$ and $x(t + \tau)$ contain different information. First, [Kaplan and Glass, 1995] suggest a suitable τ is found when the autocorrelation between $x(t)$ and $x(t + \tau)$ has fallen below $e^{-1} = 0.37$. Secondly, [Fraser and Swinney, 1986] suggest that the point at which new information is contained in a sample is reached when the mutual information, which can be considered a generalisation of the autocorrelation function (Equation 4.6), has its first minimum:

$$MI = H(x) + H(x + \tau) - H(x, x + \tau) \qquad (4.6)$$

with $H(x)$, $H(x + \tau)$ and $H(x, x + \tau)$ as defined in Equations 3.52 and 3.54.

We now have the tools in place to carry out a phase space reconstruction, and will first look at a practical robotics example.

4.2.4 Reconstructing the Attractor: Robotics Example (Obstacle Avoidance)

Figure 4.13 (right) shows part of the 26,000 data points we have of an obstacle avoiding robot's motion along the x axis. This data was obtained by logging the robot's position every 250 ms with an overhead camera.

In order to reconstruct the attractor, using Equation 4.5, we need to determine a suitable embedding lag τ and embedding dimension p.

We will select the embedding dimension p through a practical consideration. As discussed earlier, a differential drive robot such as the one used in this example has between two and three degrees of freedom, so that $p = 5$ is probably a sensible choice of embedding dimension.

To determine the embedding lag τ, we determine the point at which the autocorrelation $x(\tau)$ has fallen to approximately $e^{-1} = 0.37$ [Kaplan and Glass, 1995], or at which the mutual information MI (see Equation 4.6) has its first minimum [Fraser and Swinney, 1986].

Figure 4.6 shows autocorrelation and mutual information of $x(t)$. For $\tau = 29$ the autocorrelation has fallen below 0.37, and the mutual information has a minimum. We therefore select $\tau = 29$ and $p = 5$[1] to reconstruct the attractor, using Equation 4.5.

---

[1] $p = 5$ is a conservative, "safe" choice — see the discussion regarding the degrees of freedom of a mobile robot on page 87.

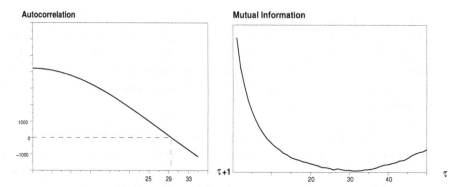

Figure 4.6. Autocorrelation (*left*) and mutual information (*right*) of $x(t)$ for obstacle avoidance behaviour. For $\tau \approx 29$ the autocorrelation has fallen below 0.37, and the mutual information has the first minimum

Finally, having reconstructed the attractor, we check whether the attractor is well "opened up" (see also page 103 for a detailed discussion). Figure 4.7 shows a reconstructed attractor for $p = 3$ (because higher dimensions can't be displayed visually), and indeed even for this lower embedding dimension the attractor is well opened up. To analyse the phase space of an obstacle avoiding robot, for example by computing Lyapunov exponent or correlation dimension (see below), we would use a higher embedding dimension, based on the practical consideration discussed above.

**Phase Space Reconstruction
(Obstacle Avoidance Behaviour)**

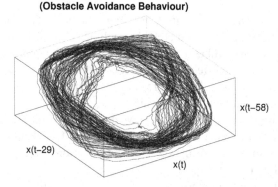

Figure 4.7. Three-dimensional reconstruction of the phase space for an obstacle avoiding robot

4.2.5 Experimental Setup and the Use of Quantitative Descriptions of Behaviour

We said earlier (Section 2.4.2) that provided we had some quantitative descriptor of a robot's behaviour — which emerges from the interaction between robot, task and environment — we could use this descriptor to characterise exactly one component of the triple robot-task-environment if we left two components unchanged, and modified the third one in a principled way, using the quantitative descriptor to characterise the relationship between changed component and observed robot behaviour.

Figure 4.8 shows such an experimental setup. Operating a Magellan Pro mobile robot in the arena shown in Figure 4.9, we conduct three experiments, by differing only one component at a time of the triple robot, task and environment.

Between experiment I and II (Figure 4.8, left and middle) we keep robot and environment constant, and change the task. Between experiment II and III (Figure 4.8, middle and right), we keep robot and task constant, and change the environment. The purpose of the experiments is to *measure* the influence of this change on the robot's behaviour.

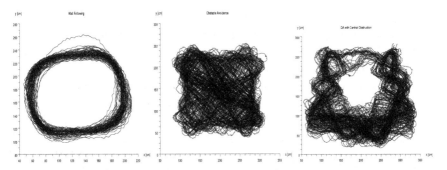

Figure 4.8. The three different data sets (robot trajectories) used in this chapter: set 1406 (wall following), set 2406 (billiard ball obstacle avoidance) and set 0507 (billiard ball obstacle avoidance with off-centre obstruction). From set 1406 to 2406 robot and environment have remained the same, but the task changed. Between set 1406 and 0507 robot and task remained the same, but the environment changed. See also Figure 1.2

4.3 Describing (Robot) Behaviour Quantitatively Through Phase Space Analysis

4.3.1 Deterministic Chaos

When asked to predict the motion of a model train, moving at a constant velocity on a circular track, one will have little difficulty and make only a very small

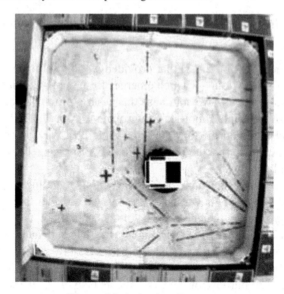

Figure 4.9. A simple robot arena

prediction error. When asked to predict the outcome of a throw of a die, one can only resort to a random guess, or always predict the mean of all numbers on the die. The former system is deterministic and fully predictable, the latter is stochastic and not predictable.

There is a third kind of system: it is deterministic rather than random, and yet not predictable, unless predictions concern the immediate, short term future. The weather falls into this category, so does the motion of a billiard ball, or the motion of a mobile robot in some cases. This kind of system is said to exhibit deterministic chaos. "Deterministic", because the system's behaviour is governed by deterministic laws such as the laws of motion, rather than randomness. "Chaotic", because it appears to behave like a stochastic system, and cannot be predicted for all times.

Specifically, deterministic chaos is said to be present if the system under investigation exhibits these four properties:

1. The system's behaviour is (predominantly) deterministic.
2. The system's behaviour is bounded and stationary.
3. The system's behaviour is sensitive to slight changes in initial conditions.
4. The system's behaviour is aperiodic.

The first two points determine whether the methods presented below are suitable for the kind of signal we have, the second two question determine whether the signal exhibits deterministic chaos or not. We are going discuss to points 1 and 2 in section 4.3.2, point 3 in section 4.4 and point 4 in section 4.5.

4.3.2 Testing for Determinism and Stationarity

Is there a Deterministic Component to the System?

All considerations presented in this chapter refer to deterministic systems, *i.e.* systems that are not mainly governed by stochastic (random) behaviour. We therefore need to establish first whether the time series $x(t)$ is deterministic, *i.e.* casually dependent on past events, or not. To do this, we use the following method, described by Kaplan and Glass [Kaplan and Glass, 1995, p. 324ff] (see also
[Kennel and Isabelle, 1992]).

The underlying assumption in determining whether the signal is deterministic or not is that in a deterministic signal D of length $2T$, the first half of the signal should be usable as a "good" predictor for the second half — in a purely stochastic (random) system this assumption would not hold. In other words: if a model-based prediction of the system is perfect (zero prediction error), the system is purely deterministic. If there is some small prediction error, the system has a deterministic component, and if the model-based prediction is only as good as a random guess, the system is not deterministic at all.

To find out whether the first half of D is a good predictor of the second half, we split the time series D into two halves of length T each, and construct an embedding $\mathbf{D}$ as given in Equation 4.7:

$$\mathbf{D}(T+i) = [D(T+i), D(T+i-1), D(T+i-2)], \forall i = 3 \ldots T \quad (4.7)$$

In other words, we construct an embedding for the second half of the time series, using an time lag τ of 1 and an embedding dimension p of 3 (of course, one could use other values for τ and p).

To make a prediction of $D(t_k+1)$ ($T < t_k \leq 2T$), we determine the closest point $\mathbf{D_c}(t_c)$ ($0 < t_c \leq T$) to $\mathbf{D}(t_k)$ in Euclidean distance, and select $D(t_c+1)$ as the prediction of $D(t_k + 1)$. In this fashion all points of the second half are predicted (we always only predict one-step ahead). Figure 4.10 shows this.

We then compute the mean squared prediction error ϵ. In order to decide whether this error is "large" or "small", we set it in relation to the error ϵ_b of a baseline prediction of simply using the average of the first half of the signal as a prediction of the second. In a purely stochastic signal the ratio ϵ/ϵ_b is 1 or larger than 1, indicating that the mean would have been the best prediction possible, and therefore that the system is non-deterministic. If, on the other hand, the ratio ϵ/ϵ_b is smaller than 1, this indicates that the first half of the time series indeed is a good predictor of the second, and that therefore the time series has a deterministic component.

There is a second way of establishing whether the time series is deterministic (*i.e.* signal values are dependent of signal values in the past) or stochastic (*i.e.*

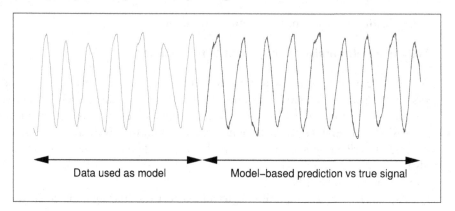

Figure 4.10. Prediction of the robot's movement along the x-axis, performing obstacle avoidance behaviour

signal values are independent from those of the past). By simply plotting $x(t)$ *vs* $x(t-\tau)$ one sees visually whether there is a causal relationship between past and present signal values, or not. These plots are called return plots, and Figure 4.11 shows three examples.

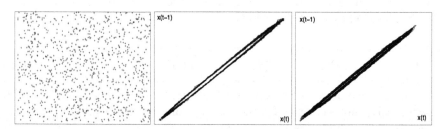

Figure 4.11. Return plots for random noise (*left*), wall following behaviour (*middle*) and obstacle avoidance behaviour (*right*). Wall following and obstacle avoidance both clearly have a deterministic element

Testing for Stationarity

The analysis of a dynamical system such as a mobile robot, interacting with its environment, is the attempt to identify and to quantify the statistical characterisations of that interaction. This means that if the statistical characterisations are not constant in time — non-stationarity — the methods for time series analysis presented in this book cannot be applied. Before analysing a signal using the methods outlined in this book, therefore, it is necessary to establish that the signal in question is stationary.

A signal is defined as stationary if it shows similar behaviour throughout its duration. "Similar behaviour" here is defined as having similar mean, standard deviation and autocorrelation structure throughout the time series [Kaplan and Glass, 1995, p.314] [Pena et al., 2001, p.29]. In practice, real world signals often show constant mean and autocorrelation structure, but different variance throughout the time series. Such signals are sometimes referred to as "weakly stationary", but considered near enough the ideal of stationarity to be treated as stationary signals. The concept of stationarity of clearly relative to the highest frequency component in the data, meaning that a meaningful analysis can only be conducted if data is used that contains the longest period inherent in the data.

To test for stationarity, therefore, entails testing whether mean and standard deviation (we do not consider the autocorrelation function for the following tests) of different parts of the data differ significantly from one another or not.

A simple test to investigate this is to divide the time series into a number of sections, and to test whether the distributions underlying each section differ from one another or not. Because it is unknown whether the underlying distributions are normal or not (this could be tested, however, using the method described in Section 3.3.2), a non-parametric test like the non-parametric ANOVA discussed in Section 3.4.4 is suitable. If the underlying distributions *are* normal, a t-test (Section 3.3.4) can be used instead of the non-parametric ANOVA.

Non-Parametric Runs Test

Another method for testing stationarity is this: As before, the data is divided into segments (for the test described below, at least 41 segments are needed). We then determine whether the mean of each segment is above or below the median of the entire time series. This will result in a sequence of, say, 41 values like "AABABBABBAA...", with "A" indicating that the mean of the segment is above the median of the entire series, and "B" that it is below.

If the time series is *not* stationary, for example because there is a linear trend upwards on the data, we will get a sequence with very few "runs" (see Section 3.5 for a definition of "run"), because all the "early" data will tend to have a mean below the median, and all the "late" data will be above the median. If, on the other hand, the data is stationary, we expect the distribution of "As" and "Bs" to be random. In other words: too few or too many runs indicate dependency between observations, and therefore non stationarity. Therefore, we test the time series in question for a random distribution of As and Bs as discussed in Section 3.5. If the distribution is random, the data is considered stationary.

One note on segmenting the data. Consider, say, a pure sine wave, which is obviously stationary. It is possible, however, to segment a sine wave so awkwardly that a sequence like "ABABABAB..." is obtained — clearly not random! The correct formulation of our procedure, therefore, should be: "If there is at least one segmentation of the time series for which randomness can be shown through the runs test, then the time series is considered stationary."

Making Non-stationary Data Stationary

Non-stationary signals can often be made stationary by simple transformations. The simplest of these transformations is to compute the first difference between successive values of the time series. If the first difference is still not stationary, the process can be repeated (second difference).

Another obvious step to take is to remove linear trends. This is achieved by simply subtracting that linear function $y = ax + b$ that best fits the time series (linear regression). This was discussed in Section 3.6.1.

If the signal $x(t)$ shows exponential growth over time, it sometimes can be made stationary by using $x(t)/x(t-1)$ for analysis [Kaplan and Glass, 1995, p.315].

Other transformations that may render non-stationary signals stationary are logarithmic or square root transformations. An exponential signal, for instance, can be linearised by computing the logarithm, and then made stationary by computing the first difference. Similarly, time series that follow a power law can be linearised by computing square roots or higher order roots.

Having established that the descriptor of the agent's behaviour (the logged time series) is indeed mainly deterministic and stationary, we are now ready to analyse the system's phase space quantitatively.

4.4 Sensitivity to Initial Conditions: The Lyapunov Exponent

Consider the following thought experiment: you observe the motion of a physical system, a pendulum, say, in physical space. That motion corresponds to a motion in phase space, the "orbit". In the pendulum's case, the orbit has the shape of an ellipse.

If you imagine starting the pendulum off at some location ϕ, and at the same time (it is a thought experiment!) at a point $\phi + \Delta$, where Δ is a small distance, we then have two motions through physical space and phase space that have started very close to each other. For some systems, like the pendulum, these two motions will neither get further apart nor closer together over time, for other systems the two orbits will converge into one orbit, and for yet other systems the two orbits will diverge and very quickly be far apart from each other (this would happen, for example, in the case of a billiard ball). The rate of divergence or convergence of two orbits that started infinitesimally close to each other describes one property of the attractor — it is known as the Lyapunov exponent.

Lyapunov Exponent and Chaos

One of the most distinctive characteristics of a chaotic system is its sensitivity to a variation in the system's variables: two trajectories in phase space that started close to each other will diverge from one another as time progresses, the more chaotic the system, the greater the divergence.

Consider some state S_o of a deterministic dynamical system and its corresponding location in phase space. As time progresses the state of the system follows a deterministic trajectory in phase space. Let another state S_1 of the system lie arbitrarily close to S_o, and follow a different trajectory, again fully deterministic. If d_o is the initial separation of these two states in phase space at time $t = 0$, then their separation d_t after t seconds can be expressed by Equation 4.8:

$$d_t = d_o e^{\lambda t} \tag{4.8}$$

Or, stated differently, consider the average logarithmic growth of an initial error E_0 (the distance $|x_0 - (x_0 + \epsilon)|$, where ϵ is some arbitrarily small value and x_0 a point in phase space) [Peitgen et al., 1992, p. 709]. If E_k is the error at time step k, and E_{k-1} the error at the previous time step, then the average logarithmic error growth can be expressed by Equation 4.9:

$$\lambda = \lim_{n \to \infty} \lim_{E_0 \to 0} \frac{1}{n} \sum_{k=1}^{n} log|\frac{E_k}{E_{k-1}}| \tag{4.9}$$

λ (which is measured in s^{-1} or in bits/s, depending on whether the natural logarithm or a logarithm to base 2 is used) is known as the Lyapunov exponent.

For an m-dimensional phase space, there are m λ values, one for each dimension. If any one or more of those components are positive, then the trajectories of nearby states diverge exponentially from each other in phase space and the system is deemed chaotic. Since any system's variables of state are subject to uncertainty, a knowledge of what state the system is in can quickly become unknown if chaos is present. The larger the positive Lyapunov exponent, the quicker knowledge about the system is lost. One only knows that the state of the system lies somewhere on one of the trajectories traced out in phase space, *i.e.*, somewhere on the attractor.

The Lyapunov exponent is one of the most useful quantitative measures of chaos, since it will reflect directly whether the system is indeed chaotic, and will quantify the degree of that chaos. Also, knowledge of the Lyapunov exponents becomes imperative for any analysis on prediction of future states.

4.4.1 Estimation of the Lyapunov Exponent of a Time Series

One method to determine the Lyapunov of an attractor describing the behaviour of a physical system is to estimate it from an observed time series of the system's motion [Peitgen et al., 1992]. However, the estimation of a Lyapunov exponent from a time series is not trivial, and often strongly dependent upon parameter settings. It is therefore not sufficient to simply take an existing software package, select parameter settings that seem appropriate, and compute the exponent. Instead, computations have to be performed for ranges of settings. There will usually be ranges of settings for which the computed Lyapunov exponent does not change — so-called scaling regions. These scaling regions indicate good parameter settings and reliable results.

There are a number of software packages available for the computation of Lyapunov exponents from a time series, for example [Kantz and Schreiber, 2003], discussed in [Kantz and Schreiber, 1997], [ANS, 2003] (discussed in [Abarbanel, 1996]) and [Wolf, 2003], based on [Wolf et al., 1995]. The results presented here were obtained using Wolf's software.

Estimating the Lyapunov Exponent, using Wolf's Program: Robotics Example

We are interested to compute the Lyapunov exponent for two different robot behaviours: wall following and obstacle avoidance. Figure 4.12 shows trajectories of a Pioneer II mobile robot executing these behaviours, logged from an overhead camera.

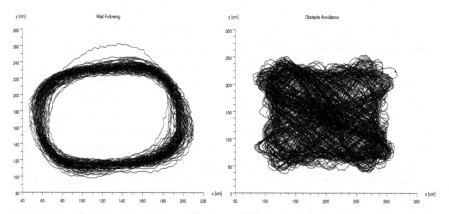

Figure 4.12. Wall following (data set 1406) and obstacle avoidance behaviour (data set 2406) of a Pioneer II mobile robot, observed from an overhead camera

In order to carry out the analysis, we look at the x and y coordinate of each behaviour individually, and estimate the Lyapunov exponent of them. Figure 4.13 shows sections of the robot's movement in x-direction for both behaviours.

Looking at the trajectories in Figure 4.12, we conjecture that the wall following behaviour is far more predictable than the obstacle avoidance one, and will

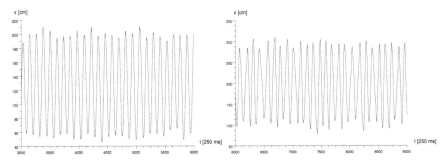

Figure 4.13. Sections of the x-axis of wall following (*left*) and obstacle avoidance behaviour (*right*)

therefore have a lower Lyapunov exponent. We use [Wolf, 2003] to investigate this.

In order to use his program, the following parameters need to be specified:

1. Embedding lag τ for the reconstruction of the underlying attractor through time-lag embedding (see Section 4.2.3). [Kaplan and Glass, 1995] suggest that a suitable value for τ is found when the autocorrelation of the time series has fallen below 1/e=0.37. Alternatively, the mutual information information can be used [Fraser and Swinney, 1986], as discussed on page 93. The general rule is: τ should be chosen such that the reconstructed attractor is properly unfolded *i.e.* contains no self-intersections) and shows a well defined structure.

2. Embedding dimension p. Takens' and Mañe's theorem [Mañe, 1981] [Takens, 1981] states that an attractor can be constructed *perfectly* through time-lag embedding, if the embedding dimension p is chosen as $p = 2d + 1$, with d being the dimension of the attractor (see Section 4.5 regarding the determination of the dimension of an attractor). In practice, however, a lower embedding dimension is usually sufficient, and for most mobile robotic systems an embedding dimension of 3 to 5 is adequate. The choice of p is limited by the amount of data available: as a rough estimate, the number of data points needed for analysis is 30^d [Wolf, 2003], where d is the dimension of the attractor (see Section 4.5).

   A well established method to determine a suitable embedding dimension is the method of false nearest neighbours. This is discussed in [Kennel et al., 1992].

3. Time step. This is simply the sampling rate used for obtaining the time series (in the examples used here the overhead camera logged the robot position every 250ms).

4. Evolution time *evolv*. This parameter determines how many steps a pair of points will be followed through phase space, to estimate their divergence

over time. The right setting for this (typically 3 to 12) must be determined through finding a scaling region.

5. Minimum separation at replacement. This indicates the minimum separation between two points in phase space to be selected as a pair of points that is traced through phase space. As a rule of thumb, this value should be set to zero for noise-free data, and to 1-2% of the range of time series values for noisy data.

6. Maximum separation $maxdist$ for replacement. This indicates the maximum allowable separation between points in phase space before a new pair is sought to estimate divergence. As a rule of thumb, this value can be set to 10-15% of the range of time series values for noisy data, but suitable values need to be determined by finding a scaling region.

Bearing these considerations in mind, we can now estimate the Lyapunov exponents for the wall following and the obstacle avoidance behaviour.

Lyapunov Exponent of the Wall Following Behaviour

For the wall following behaviour, which consists of 13,000 data points (54 min of robot operation), at $\tau = 30$ and a minimum separation of 2 (the range of the x-axis time series is 164), we obtain the results shown in Table 4.1.

Table 4.1. Estimations of the Lyapunov exponent for the wall following behaviour, for a range of different parameter settings, using Wolf's program [Wolf, 2003]

$evolv$	$maxdist$	$\lambda(p=3)$	$\lambda(p=4)$	$\lambda(p=5)$
2	20	0.005	0.004	0.003
3	20	0.006	0.02	0.03
4	20	0.03	0.03	0.03
5	20	0.03	0.03	0.03
6	20	0.04	0.03	0.03
7	20	0.04	0.04	0.03
8	20	0.03	0.04	0.03
9	20	0.04	0.03	0.03
10	20	0.04	0.03	0.03
6	15		0.05	
6	20	0.04	0.03	0.03
6	25		0.02	
6	30		0.02	
6	35		0.02	

Results are fairly uniform throughout, and taking an evolution time of 6 and an embedding dimension of 4, we estimate $0.02 < \lambda_{wf} < 0.03$ bits/s.

Lyapunov Exponent of the Obstacle Avoidance Behaviour

For the obstacle avoidance behaviour (26,000 data points, 108 min of operation, range 239), using a $\tau = 29$ (see Figure 4.6) and a minimum separation of 2 we obtain the results given in Table 4.2.

Table 4.2. Estimations of the Lyapunov exponent for the obstacle avoidance behaviour, for a range of different parameter settings, using Wolf's program [Wolf, 2003]

evolv	Max.dist.	$\lambda(p=3)$	$\lambda\,(p=4)$	$\lambda\,(p=5)$
2	28	0.02	0.02	0.01
3	28	0.14	0.12	0.10
4	28	0.15	0.12	0.11
5	28	0.15	0.12	0.11
6	28	0.15	0.12	0.11
7	28	0.14	0.12	0.10
8	28	0.15	0.12	0.10
9	28	0.14	0.12	0.10
10	28	0.14	0.12	0.10
6	20	0.18	0.14	0.10
6	25	0.16	0.13	0.11
6	28	0.15	0.12	0.11
6	30	0.14	0.11	0.10
6	35	0.13	0.11	0.10
6	40	0.11	0.10	0.09

We can see a scaling region for $3 < evolv < 10$, and therefore select $evolv=6$. The maximum separation at replacement doesn't show such a pronounced scaling region, but for values between 25 and 30 results are fairly consistent. The results for $p = 4$ and $p = 5$ are in good agreement, and we select $p = 4$. From this, we estimate $0.11 < \lambda_{oa} < 0.13$ bits/s. Obstacle avoidance behaviour indeed is more chaotic and has a higher Lyapunov exponent than wall following behaviour, as hypothesised earlier.

4.4.2 Estimating the Lyapunov Exponent Through Information Loss and Prediction Horizon

Information Loss

As the Lyapunov exponent is expressed as information loss per time unit, it can be estimated by determining how much information about the system is available at time t_0, and how this information decreases with increasing t. This can be achieved in the following manner.

Let's assume we observe a mobile robot interacting with its environment, logging the robot's position $x(t)$ and $y(t)$. The information $I(t)$ (in bits) we have regarding the robot's position is then given by

$$I(t) = lnB(t)/ln2 \qquad (4.10)$$

where $B(t)$ is the number of distinct locations along the x and y axes we can identify at time t. If, for example, we can pinpoint the robot's position on the x axis to one out of 256 distinct locations (*e.g.* one camera pixel out of 256), the information available is 8 bit.

The way we identify the robot's position in this thought experiment is by *predicting* it, based on past data, in the same way described in Section 4.3.2. Initially, the robot's position is perfectly known, and is limited only by the resolution of our camera logging system.

Let us assume that we use an overhead camera to log the position of the robot. Let us also assume that at time $t = 0$ we know the robot's position to an accuracy of one pixel. In order to compute the information at time $t = 0$, therefore, we can simply use the range R, defined as $R = x_{max} - x_{min}$, with x_{max} and x_{min} being the largest and smallest x position observed, and compute $I(t = 0)$ using $B = 1$ (Equation 4.10), because at $t = 0$ we are able to specify the robot's position to an accuracy of $1/R$ pixels.

As time progresses, we become increasingly less able to predict the robot's position, due to increasing prediction error. A prediction error of 1 at some time t_1, for example, means that we are now only able to specify the robot's position on the x axis to an accuracy of $2/R$ pixels — the identified pixel location $\pm$ one pixel.

Say we had initially 256 distinct positions at time t_0. This means $I_0 = 8$ bit. For a prediction error of 1 at time t_1, we would be able to localise the robot as one out of 128 distinct positions (7 bit). In other words, it has taken $t_1 - t_0$ seconds to lose 1 bit of information, which gives us an estimate of the Lyapunov exponent as $\lambda \approx \frac{1bit}{(t_1 - t_0)s}$.

The Prediction Horizon

The Lyapunov exponent, whose unit is bits/s, indicates the loss of information due to the chaotic nature of the signal as one predicts the signal for longer and longer times ahead. A perfectly noise-free and non-chaotic signal, with a Lyapunov exponent of zero, can be predicted for any length of time, without suffering from a prediction error. On the other hand, a chaotic signal cannot be predicted for arbitrary lengths of time, because with each prediction step uncertainty increases, until finally the prediction is no longer better than an educated guess. At this point "complete loss of predictability" has occurred.

If, for example, you are asked to predict the temperature in your home town 10 min ahead and to give an uncertainty indication of your prediction, you can make a fairly good prediction with a small uncertainty interval, by simply saying the temperature is going to be the same as it is at the moment. To predict 2 h ahead, you will be a little less certain, even more so for 12 h ahead, and eventually your prediction will on average be no better than an educated guess (*e.g.* looking up the mean temperature in a tourist guide).

The Lyapunov exponent can be used to compute when this complete loss of predictability will occur, *i.e.* when any model of your data is going to perform no better than an educated guess (we refer to this point in time as the "prediction horizon"). Bear in mind that the Lyapunov exponent is an averaged measure — there may well be situations in which predictions are better than educated guesses well beyond the estimated prediction horizon, but on average the prediction horizon estimated by the Lyapunov exponent is when complete loss of predictability occurs.

By way of illustration, let's assume you are measuring the pressure in some industrial plant, and you would like to predict what the pressure is going to be at some time in the future. Having logged a sufficiently long time series of pressure measurements in the past, you estimate the Lyapunov exponent to be 0.5 bit/s. The pressure sensor you are using has a resolution of 256 different pressure values, *i.e.* log(256)/log(2)=8 bit. This means that on average total loss of predictability will happen after 16 s. In other words: on average even a "gold standard" model of the pressure profile will do no better than an educated guess of the pressure after 16 s.

"Educated guess" here means a prediction of a value that is not based on specific past values, but that exploits global properties of the signal. It is the baseline against which we compare the prediction performance of our model (which does take past values into account). A simple baseline to use would be for each point $x(t_p)$ whose development over time we would like to predict to pick some other point $x(t_m)$ randomly from the time series, and to use the successors of $x(t_m)$ as predictions of the successors of $x(t_p)$.

As the Lyapunov exponent can be used to estimate that point in time at which an educated guess will produce as small a prediction error (on average) as a "gold standard" model, we should be able to do the reverse as well: determine that point in time at which we might as well make random guesses about the signal, and deduce the Lyapunov exponent from this.

The "gold standard" model we will use is the data itself. Splitting the data into two equal halves of length T each, we will use the first half of the data as a model of the second. This is a sensible procedure, since we only deal with deterministic data here, meaning that past data points are to some degree predictive of future data points.

In order to predict future data points $D(t_2), t_2 = T + 1 \ldots 2T$ of the second half of our data, we construct a three-dimensional embedding $\mathbf{D}(t_2)$ as given in Equation 4.11, and search through the first half of the data for the vector $\mathbf{D}(t_1), 1 \leq t_1 \leq T$ that is closest to $D(t_2)$ (Euclidean distance):

$$\mathbf{D}(t_2) = [D(t_2), D(t_2 - \tau), D(t_2 - 2\tau)] \qquad (4.11)$$

with τ the embedding lag, as described on page 103. This is a three-dimensional reconstruction of phase space, by now familiar to us.

We then predict the next k data points $D_m(t_2 + 1) \ldots D_m(t_2 + k)$ as given in Equation 4.12:

$$D_m(t_2 + i) = D(t_1 + i), i = 1 \ldots k, 1 < t_1 < T, T < t_2 < 2T \quad (4.12)$$

This prediction we will compare against our baseline, which states that we select a point $D_B(t_r)$, $1 < t_r < T$ at random, and predict $D_B(t_2 + i)$ as given in Equation 4.13:

$$D_B(t_2 + i) = D_B(t_r + i), i = 1 \ldots k, 1 < t_r < T, T < t_2 < 2T \quad (4.13)$$

The point at which the average model error $\overline{D_M - D}$ is the same as the average baseline error $\overline{D_B - D}$ is the prediction horizon.

4.4.3 Estimating the Lyapunov Exponent Using Information Loss and Prediction Horizon: Standard Examples

Before we apply the two methods discussed above to examples from robotics, we test them, using equations whose chaotic properties are known. Specifically, we estimate the Lyapunov exponent for the quadratic iterator and the Lorenz attractor, two well understood systems. This serves as a check to confirm that the proposed methods work for known chaotic systems.

Estimating the Lyapunov Exponent of the Quadratic Iterator

The quadratic iterator (also known as the logistic map) is a well known, chaotic dynamical system, given by Equation 4.14:

$$x(t) = 4x(t-1)(1 - x(t-1)) \quad (4.14)$$

A typical time series of the quadratic iterator is shown in Figure 4.14.

Figure 4.15 shows the computed information loss for the quadratic iterator, as predictions are made for increasingly longer times ahead. There is a linear region for the information loss between seconds 1 and 7, during which interval 6 bits of information are lost. This means that $\lambda = 1$ bit/s for the quadratic iterator.

The estimate of the prediction horizon is shown in Figure 4.16. It indicates that the prediction horizon — the point at which all information is lost — is reached after 10 s. As there were 10 bits of information available initially (see Figure 4.15), this means a Lyapunov exponent of $\lambda = 1$ bit/s, the same result as the result obtained using the information loss. Incidentally, the correct Lyapunov exponent of the quadratic iterator is indeed $\lambda = 1$ bit/s [Peitgen et al., 1992], our results are therefore in order.

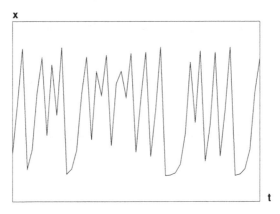

Figure 4.14. Time series obtained through Equation 4.14

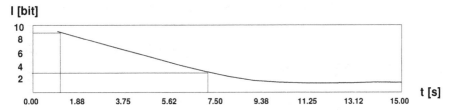

Figure 4.15. Information loss for the quadratic iterator. A linear region is visible between seconds 1 and 7, in those 6 s 6 bits of information are lost

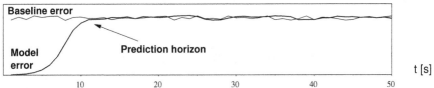

Figure 4.16. Prediction horizon for the quadratic iterator shown in Figure 4.14

Estimating the Lyapunov Exponent of the Lorenz Attractor

The Lorenz attractor is another well known dynamical system that has chaotic properties, it is defined by the differential Equations 4.15.

$$\dot{x} = 16x + 16y$$
$$\dot{y} = x(45.92 - z) - y$$
$$\dot{z} = -4z + xy \tag{4.15}$$

Figure 4.17 shows the solution of Equation 4.15 for $x(t)$, it is this time series that we will use to estimate λ_{Lorenz}.

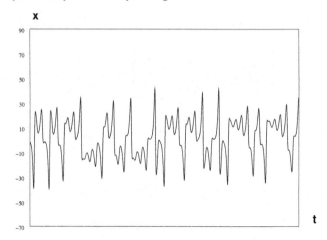

Figure 4.17. The solution of Equation 4.15 for $x(t)$

Figure 4.18 shows the estimated information loss and prediction horizon for the Lorenz attractor.

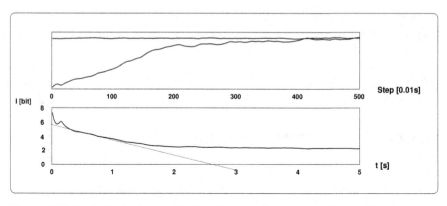

Figure 4.18. Prediction horizon and information loss for the x coordinate of the Lorenz attractor (see Figure 4.17)

The prediction horizon is about 400 steps, which, owing to the Runge-Kutta method used to solve Equation 4.15, equates to 4 s. As the initial information available was approximately 7.5 bit, we estimate $\lambda_{Lorenz} \approx \frac{7.5bit}{4s} = 1.9$ bit/s, using the prediction horizon.

The information loss (Figure 4.18, bottom) shows a linear region which is indicated in the graph. The line indicates that within 2.6 s the available information decreases from about 5.6 bits to zero bits, resulting in an estimated Lyapunov exponent of $\lambda \approx \frac{5.6}{2.6} = 2.15$ bit/s.

We therefore estimate 1.9 bits/s $< \lambda_{Lorenz} < 2.15$ bits/s. The accurate value is $\lambda_{Lorenz} = 2.16$ bit/s [Wolf et al., 1995], a good agreement with our results.

4.4.4 Estimating the Lyapunov Exponent Using Information Loss and Prediction Horizon: Robotics Examples

Wall Following

Figure 4.19 shows prediction horizon and information loss for the robot's wall following behaviour.

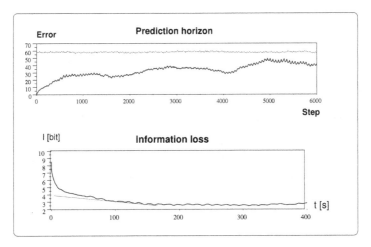

Figure 4.19. Prediction error and information loss for the wall following behaviour (data set 1406)

From Figure 4.19 we can see that in the case of the (highly predictable) wall following behaviour the prediction horizon is very long indeed. Wall following is so repetitive and predictable that even for predictions of 6000 steps ahead (more than 25 min at a sampling rate of 4 Hz) we are able to predict the robot's position more accurately, using the robot's past behaviour as our model, than just randomly guessing. From this consideration it follows that the estimated Lyapunov exponent is about zero.

Figure 4.19 also shows that the information loss is only linear for a limited region of t, indicating that we only get an exponential information loss, as hypothesised by Equation 4.8, for a limited period of time.

If any linear region can be identified in the graph for the information loss at all (Figure 4.19, bottom), it indicates an information loss of about 4 bits in 200s, which yields $\lambda_{wf} \approx 0.02$ bits/s, *i.e.* essentially zero. These results are in very good agreement with the results obtained using Wolf's method (Table 4.1).

Obstacle Avoidance

We can apply the same considerations to the obstacle avoidance behaviour shown in Figure 4.12. As before, we split the data into two halves, and use the first half to predict the second, reconstructing a three-dimensional phase space as given in Equation 4.11. The results are shown in Figure 4.20.

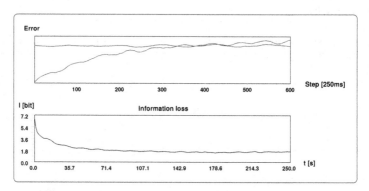

Figure 4.20. Prediction error and information loss for the obstacle avoidance behaviour (data set 2406)

Figure 4.20 reveals that unlike in the case of wall following, which is a highly predictable behaviour, here the prediction horizon is much shorter, about 320 steps (80 s). In other words: we reach the prediction horizon after 80 s, beyond which model-based predictions are as good as taking a randomly selected data point as a predictor.

We can estimate the Lyapunov exponent of the obstacle avoidance behaviour from Figure 4.20: As the initially available information was about 7.9 bit and the prediction horizon is 80s, we estimate $\lambda \approx 7.9$ bit/80s = 0.1 bit/s. This is in good agreement with the results obtained earlier, using Wolf's method (Table 4.2).

This has interesting implications for instance for computer modelling of our robot's obstacle avoidance behaviour: However good a model, it cannot be able to predict the *exact* trajectory of the robot further than about 80 s ahead (however, the model may still be able to model the underlying dynamical properties of the robot's interaction with its environment).

Random Walk

Figure 4.21 shows the trajectory of a third example, random walk obstacle avoidance in a slightly more complex environment.

Computing information loss and prediction horizon for this example yields the results shown in Figure 4.22. After about 80 s there is no significant difference between the baseline error and the model error, *i.e.* the prediction horizon in this case is 80 s.

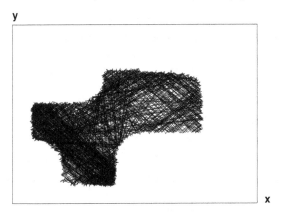

Figure 4.21. Random walk obstacle avoidance

The range of x-positions logged in this case is 226. As we are able to log the robot's position with an accuracy of 1 pixel, this means that the information I_0 that is initially available about the robot's location is $I_0 = ln\frac{226}{1}/ln2 = 7.8$ bits. We therefore estimate $\lambda_{rw} \approx \frac{7.8bits}{80s} = 0.1$ bits/s.

The information loss (Figure 4.22, bottom) does not show a pronounced linear region, indicating that the exponential model of information loss does not fit this data well.

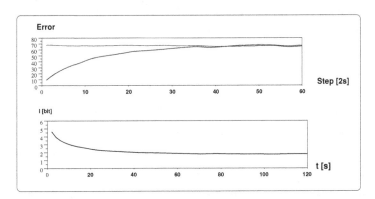

Figure 4.22. Prediction horizon and information loss for random walk obstacle avoidance

4.4.5 Estimating the Lyapunov Exponent from Prediction Horizon and Attractor Size

There is an alternative way of relating the Lyapunov exponent to "the real world", again using the robotics example of wall following and obstacle avoidance behaviour.

The Lyapunov exponent λ was defined earlier (page 100) as $D(t) = D_0 e^{\lambda t}$, where $D(t)$ is the separation between two points in phase space at time t, and D_0 some small initial separation between two points in phase space at time $t = 0$. Arguably, we have lost all information about the robot's position when all we can say about the robot's position is that it is "somewhere on the attractor". Obviously, at this point t_{crit} our initial error D_0 has increased to the size of the entire attractor S, so that Equation 4.16 holds:

$$D_0 e^{\lambda t_{crit}} = S \tag{4.16}$$

We can use the following simplifying assumptions to estimate the Lyapunov exponent from graphs such as Figures 4.19 and 4.20. Let's assume all N points in phase space are evenly distributed, and have a separation of D_0. The number n of points along each dimension of the d-dimensional attractor is then given by Equation 4.17:

$$N = n^d \tag{4.17}$$

with d being the dimension of the attractor.

The side length S of the attractor is consequently given by Equation 4.18:

$$S = D_0 n^d = N D_0 \tag{4.18}$$

From Equations 4.16 and 4.18 follows Equation 4.19. Equation 4.19 allows us to estimate the Lyapunov exponent from t_{crit}, the time at which our model is only able to make random guesses, rather than informed predictions:

$$\begin{aligned} D_0 e^{\lambda t_{crit}} &= N D_0 \\ \lambda t_{crit} &= ln N \\ \lambda &= \frac{ln N}{t_{crit}} \end{aligned} \tag{4.19}$$

Applying Equation 4.19 to the wall following behaviour, we see in Figure 4.19 that the prediction error of our data-driven model essentially never flattens out, indicating a prediction horizon of infinity (in reality, the prediction horizon is going to be a number larger than 6000). This means that the estimated Lyapunov exponent is $\lambda \approx 0$, which is in agreement with our earlier computations.

The error curve of the data-model driven prediction for the obstacle avoidance behaviour (Figure 4.20) approaches the baseline error after about 600 prediction steps ($t_{crit} = 150s$). We have $N = 26,000$ data points here, resulting in $\lambda_{oa} = \frac{ln 26000}{150s} = 0.07/s = 0.1$ bits/s. This is in good agreement with our earlier estimate, using Wolf's method (Table 4.2).

An alternative description of this methods was contributed by my colleague Keith Walker, based on [Baker and Gollub, 1996, p.154f].

If the embedding dimension of a reconstructed attractor is m, then a Lyapunov exponent is associated with each of the m principal axes. If one considers a very small hypersphere of points in an m-dimensional phase space and track the trajectories of those points with time, the original hypersphere of points will evolve to a shape determined by the Lyapunov exponents. This occurs since a negative Lyapunov exponent associated with a principal axis will result in that dimensional component of the hypersphere contracting exponentially with time while a positive exponent will result in exponential expansion. The hypervolume containing the trajectories at time t is expressed as $V(t) = V_0 e^{(\lambda_1 + \lambda_2 + ... \lambda_m)t}$, where V_0 is the hypersphere of points at $t = 0$. Hence, for a system with a single positive Lyapunov exponent, the initial hypersphere will evolve into an ellipsoid. (Note: if two positive Lyapunov exponents exist, the system is cited as hyper chaotic.)

For dissipative systems the relationship $\sum \lambda_i < 0$ holds, and the hypervolume approaches zero with time. If one of the Lyapunov exponents is positive (for example λ_k) and all other exponents are negative, the resulting ellipsoid becomes a fine filament along the k-principal axis. This ever-thinning filament will eventually trace out most of the attractor. If the initial dimension of the hypersphere in the k-direction is d_k, then the elongation of the filament is $d = d_k e^{\lambda_k t}$. Since the attractor is bounded, the filament will never exist beyond the dimension of the attractor, D. Let t_{crit} be the time when d_k will have expanded to D. What this means is that two points in phase space that are initially separated along the k-principal axis by d_k, will now be separated by D. It is at this time that the state of the system is essentially indeterminate, or lost, and the only information known about the system is that it lies somewhere on the attractor and $\lambda_k = \frac{1}{t_{crit}} ln \frac{D}{d_k}$.

A fairly decent prediction of λ_k can therefore be made, knowing t_{crit} and utilising a simplifying assumption suggested by [Baker and Gollub, 1996]. If there are N measurements of some parameter of the system with which one reconstructs the attractor, then the mean distance along any axis between two adjacent points in phase space can be approximated by $\frac{D}{N}$. If we assign this ratio as the initial uncertainty d_k between two points in phase space, then

$$\lambda_k \approx \frac{1}{t_{crit}} ln N. \tag{4.20}$$

This can be further simplified to a rule of thumb, if one considers that $ln N$ varies slowly with N. For example: for values of $N = 1000, 10,000, 50,000, 100,000, ln N$ is 7, 9, 10.8, 11.5 respectively. Since most data consists of 1000 or more measurements one can roughly say that

$$\lambda_k t_{crit} \approx 10. \tag{4.21}$$

Applying this rule of thumb to the examples of wall following and obstacle avoidance (page 104), we get the following results. For the wall following behaviour, which had a prediction horizon of essentially infinity, t_{crit} is also infinite, resulting in an estimated $\lambda_{wf} \approx 0$ (Equation 4.21). This is in agreement with the result we got using Wolf's method.

For the obstacle avoidance behaviour, we estimated $t_{crit} \approx 150s$, resulting in an estimate of $\lambda = 0.07/s = 0.1$ bit/s. This is in agreement with results obtained earlier.

4.5 Aperiodicity: The Dimension of Attractors

Another main characteristic of a dynamical system exhibiting deterministic chaos is that the state variables never return to their exact previous values, *i.e.* the system's behaviour is not periodic. The trajectory in phase space lies on an attractor with a fractal dimension, a "strange" attractor. There is, however, variation from system to system in how close state variables return to previous values, and it is therefore desirable to quantify this degree of "proximity".

The measure to quantify the degree of aperiodicity is the correlation dimension d of the attractor. The correlation dimension indicates whether data is aperiodic or not, and to what degree: Periodic data has a correlation dimension of zero, chaotic attractors have a non-integer correlation dimension [Kaplan and Glass, 1995, p. 321].

Determining the Correlation Dimension

The dimensionality of an attractor is related to its aperiodicity: the more aperiodic the dynamics, the greater the dimension of the attractor. In order to measure how periodic a trajectory through phase space is, one uses the following idea.

Suppose you take an arbitrary point on the attractor, draw a hypersphere of radius r — the so-called "correlation distance" — around that point, and count how many points of the attractor lie within that hypersphere. This number of points is referred to as the "correlation integral" $C(r)$, given by Equation 4.22:

$$C(r) = \frac{\theta}{N(N-1)} \tag{4.22}$$

where θ is the number of times that $|\mathbf{D}(t_i) - \mathbf{D}(t_j)| < r$. i and j are two different times at which an embedding $\mathbf{D}$ is taken (Equation 4.5), and r is the "correlation distance". $N(N-1)$ is obviously the maximum number of cases where $|\mathbf{D}(t_i) - \mathbf{D}(t_j)| < r$ is theoretically possible (the trivial case $i = j$ is excluded).

In a perfectly periodic attractor, for example in the case of the ideal pendulum, the correlation integral is not going to increase with increasing r. The slope

$C(r)$ vs r is zero. In other cases, $C(r)$ is going to increase as one increases r. It is the slope of $C(r)$ vs r that is defined as the "correlation dimension" of the attractor.

In practical computations, this slope is often estimated using Equation 4.23 [Kaplan and Glass, 1995, p. 354]:

$$d = \frac{log\, C(r_1) - log\, C(r_2)}{log\, r_1 - log\, r_2} \tag{4.23}$$

where r_1 is chosen such that r_1 is roughly $\sigma/4$ (σ being the standard deviation of the time series), and $C(r_1)/C(r_2) \approx 5$ [Theiler and Lookman, 1993].

Clearly, the computation of the correlation dimension is dependent upon the chosen embedding dimension p and the correlation distance r. To compute both p and d from the same process is an ill-defined problem, and the goal is to find a range of parameters p and r for which d is computed virtually identically (a so-called "scaling region"). In other words, one aims to find a region where the computation of the correlation dimension d, using Equation 4.23, is not critically dependent upon the choice of embedding dimension p and correlation distance r.

To find such a scaling region, one can plot the correlation dimension d as a function of correlation distance r for all embedding dimensions p between, say, 1 and 10 [Kaplan and Glass, 1995, p. 323], and check whether there are regions for which the choice of r and p does not alter the computed correlation dimension d. That d is then our estimate of the dimension of the attractor.

4.5.1 Correlation Dimension: Robotics Examples

Obstacle Avoidance

In the following, we will estimate the correlation dimension of the attractor underlying the obstacle avoidance behaviour shown in Figure 4.12 (*right*).

We first compute $\frac{dC(r)}{dr} = d$, using Equation 4.22. Figure 4.23 shows the result.

Choosing a very large correlation distance r is somewhat like looking at an object from a great distance: it will have dimension zero [Kaplan and Glass, 1995, p. 323]. The other extreme, choosing too small an r, will result in signal noise being amplified, which is equally undesirable.

Figure 4.23 reveals that a scaling region — a region where d does not change when parameters are changed — exists around $r = 40$; increasing the embedding dimension in this region no longer changes the computed correlation dimension.

To obtain another representation of this result, we now fix $r = 40$ and plot d vs p (Figure 4.24). Figure 4.24 shows a "levelling off" of the computed correlation dimension d for an embedding dimension between 6 and 10. The computed

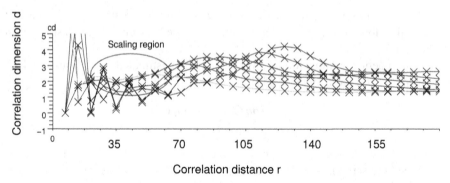

Figure 4.23. Correlation dimension *vs* correlation distance for the obstacle-avoidance behaviour shown in Figure 4.12 (*right*), for embedding dimensions 5, 7, 9, 11, 13, 15 and various correlation distances

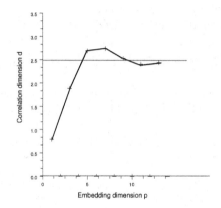

Figure 4.24. Correlation dimension *vs* embedding dimension for the obstacle avoidance behaviour shown in Figure 4.12 (*right*), for $r = 40$. The computed correlation dimension is $d \approx 2.5$.

correlation dimension d at this point is $d \approx 2.5$, a fractal dimension indicating that the attractor is strange and that the system's behaviour is aperiodic, one of the characteristics of deterministic chaos.

Wall Following

Figure 4.25 shows the correlation dimension d *vs* p and r for the wall following behaviour. We estimate $d \approx 1.4 - 1.6$.

Random Walk

Table 4.3 shows results obtained when we apply Equation 4.23 to the random walker introduced in Section 4.4.4.

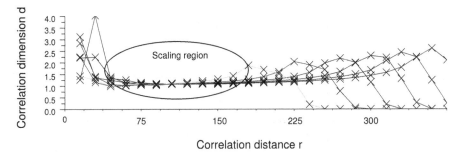

Figure 4.25. Correlation dimension *vs* embedding dimension for the wall following behaviour shown in Figure 4.12 (*left*), for embedding dimensions 3, 5, 7, 9, 11, 13 and 15. The computed correlation dimension is $d \approx 1.4$-1.6.

Table 4.3. Estimates of d for the random walker, using different embedding dimensions p and embedding correlation distances r. Equation 4.23 was used for the computation. Embedding lag $\tau = 40$

p	r_1	r_2	$C(r_1)$	$C(r_2)$	d
3	15	12	0.0093	0.0058	2.12
		10		0.0039	2.14
		7		0.0018	2.15
4	15	8	0.0035	0.00072	2.52
		5		0.00022	2.54
5	20	10	0.0032	0.00048	2.74
6	15.8	20.7	0.0027	0.0066	3.31
8	15.8	17.3	0.00094	0.0012	2.69

From Table 4.3 we estimate the dimension of the attractor of the random walker to be between 2.7 and 3.3.

4.6 Summary

The motion of a mobile robot is a function of time, and can be described, for example, through differential equations (*e.g.* speed $\dot{x}$). The robot is therefore a dynamical system, and in this chapter methods from dynamical systems theory have been applied to describe and analyse robot behaviour.

The motion of a dynamical system through physical space can be fully described by the system's motion through phase space, the space defined by the system's position $x(t)$ and speed $\dot{x}(t)$ along every of its degrees of freedom. Because there are established tools available for phase space analysis, this is an attractive option.

The phase space of a dynamical system can be reconstructed through time lag embedding (Section 4.2.3), using the observed values of a variable that is relevant to the system's operation. In the case of a mobile robot, the robot's trajectory is usually one of the most relevant parameters characterising the robot's behaviour, and the robot's phase space can be reconstructed for example by using its $x(t)$ or its $y(t)$ positions.

Once the attractor is reconstructed, it can be characterised quantitatively, for example by these three quantitative measures:

1. Lyapunov exponent (Section 4.4). This measures the information loss in bits/s, and thus establishes how quickly small initial perturbations are amplified so much that no other statement can be made about the system than "it is somewhere on the attractor", which in the mobile robotics case translates to "the robot is going to be somewhere in the arena, but it is impossible to say where."
2. Prediction horizon (Section 4.4.2). This measure is related to the Lyapunov exponent, and states the time after which a model-based prediction regarding the system's future states is, on average, as precise as a random guess.
3. Correlation dimension (Section 4.5). This is a measure of the system's periodicity: does the system ever revisit states, or does the motion through phase space merely pass through the vicinity of previously visited states?

Besides providing useful information about the mobile robot's interaction with the environment — information such as "how far ahead into the future could the best simulator possibly predict the robot's position?" — these quantitative descriptors are needed in a science of mobile robotics for another reason. The behaviour of a robot emerges from the interaction between the robot, the task (control program) and the environment. If quantitative descriptions of the robot's behaviour, such as Lyapunov exponent, prediction horizon or correlation dimension are available, a new method of robot experimentation becomes available to the researcher: two of the three components can be left unchanged, and the quantitative measure be used to investigate systematically how the third component influences the robot's behaviour. Robotics research would involve *quantitative*, rather than *qualitative* assessment, and allow independent replication and verification of experimental results, hallmarks of a maturing science.

5

Analysis of Agent Behaviour — Case Studies

Summary. In this section the techniques presented in Chapter 4 are applied to "real world" experimental data. Three data sets will be analysed: the movement of a mobile robot that is randomly moving around in its environment, a "chaos walker", and the flight path of a carrier pigeon.

5.1 Analysing the Movement of a Random-Walk Mobile Robot

Figure 5.1 shows the trajectory of a Magellan Pro mobile robot that has moved in an irregularly shaped environment for just under three hours.

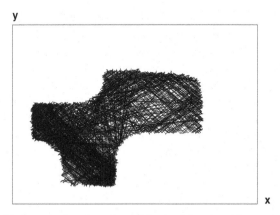

Figure 5.1. Trajectory of an obstacle-avoiding mobile robot (observation time: 3 h)

In Figure 5.2 $x(t)$ and $y(t)$ are shown separately.

We are interested in analysing this robot behaviour: is it mainly deterministic or stochastic; therefore, is it predictable, and if yes, for how many steps ahead? How can this behaviour be described quantitatively?

121

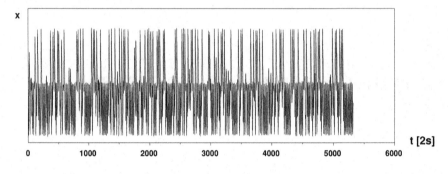

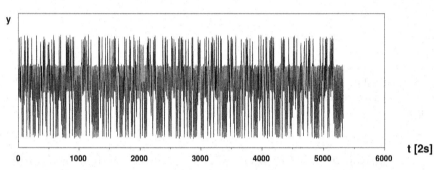

Figure 5.2. x (*top*) and y (*bottom*) *vs* time of the trajectory shown in Figure 5.1

5.1.1 Determinism

To establish whether $x(t)$ (Figure 5.2) is mainly deterministic or not, we'll apply the techniques discussed in Section 4.3.2.

The return plot of $x(t)$ *vs* $x(t-3)$ indicates that $x(t)$ is deterministic, rather than stochastic (Figure 5.3 — compare with Figure 4.11).

This result is confirmed by using a three-dimensional embedding of the first half of $x(t)$ (Equation 4.7) as a predictor of the second half, and comparing the prediction error ϵ_{model} with the baseline prediction of error ϵ_b obtained when the mean of the signal is used as a prediction. The ratio of $\epsilon_{model}/\epsilon_{mean}$ turns out to be 0.048, that is very small compared with 1.0, confirming that indeed $x(t)$ is deterministic.

5.1.2 Stationarity

Next, we will establish whether $x(t)$ is stationary or not, using the runs test described in Section 4.3.2. Dividing $x(t)$ into 110 bins, a more or less arbitrary choice, we determine whether in each bin the median value is above or below the mean of the entire series $x(t)$.

x(t−1)

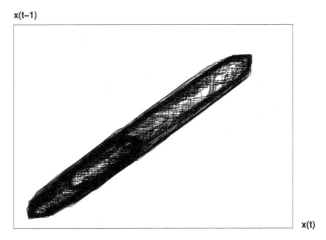

x(t)

Figure 5.3. Return plot for $x(t)$ for the random walker

It turns out that there are 55 runs, which is exactly inside the acceptance interval [45,65], indicating that the distribution of "above mean" and "below mean" medians is random: $x(t)$ is stationary.

The result that $x(t)$ is stationary is confirmed through a non-parametric analysis of variance. There is no significant difference between $x(1 - 3400s)$, $x(3400 - 6800s)$ and $x(6800 - 10400s)$ (p=0.37).

5.1.3 Predictability and Lyapunov Exponent of Random-Walk Obstacle Avoidance

Having established that our data is amenable to the mechanisms discussed in Chapter 4, we will now try to make some quantitative descriptions of the data. We'll start by looking at the predictability of $x(t)$, as discussed in Section 4.4.2.

In order to determine the prediction horizon of $x(t)$, we need to determine the correct embedding lag for reconstructing the attractor, which can be done by establishing when the autocorrelation reaches $1/e$ [Kaplan and Glass, 1995], or when the mutual information has its first minimum [Abarbanel, 1996]. Both autocorrelation and mutual information of $x(t)$ are shown in Figure 5.4.

The autocorrelation falls below e^{-1} for $\tau = 9$, the mutual information has its first minimum for $\tau = 7$. We select an embedding lag of 8.

We now predict the second half of $x(t)$, using the first half as our model, by constructing for each point $x(t)$ a three-dimensional embedding $D = [x(t), x(t-8), x(t - 16)]$, and finding the nearest neighbour D_n to each D within the first half of our data. The successors of D_n are then used to predict the successors to $x(t)$.

Figure 5.5 shows the result. We can see that for predictions of more than 35 steps ahead (which, at a data logging frequency of 0.5 Hz corresponds to 70 s) the

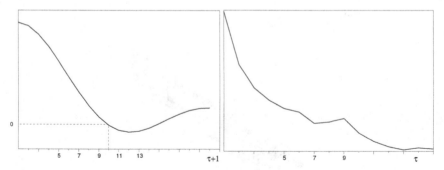

Figure 5.4. Autocorrelation (*left*) and mutual information (*right*) of $x(t)$ for the random walker

average prediction error using our data as model is as big as the prediction error obtained when using a randomly selected point from the first half as a predictor for the second half of our data. In other words: if you wanted to predict the precise location of the random-walk mobile robot, whose trajectory is shown in Figure 5.1, then this could, on average, only be done better than random guessing up to a prediction horizon of about 70 s ahead.

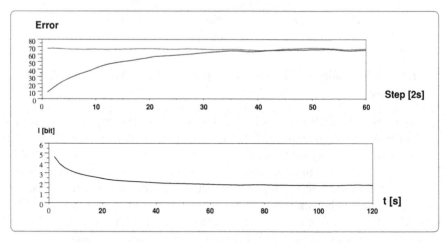

Figure 5.5. Prediction horizon and information loss for $x(t)$ for random walk obstacle avoidance

Figure 5.5 also gives us an estimate of the Lyapunov exponent of $x(t)$. The prediction horizon is 70s, as we established above. Because we initially have 5.2 bits of information (bottom graph of Figure 5.5), we estimate the Lyapunov exponent as $\lambda \approx \frac{5.2 bit}{70 s} = 0.07$ bit/s.

Using Wolf's method [Wolf, 2003], we obtain an estimate of $0.07 < \lambda < 0.1$ bit/s, which confirms our result.

5.1.4 Analysis of the Attractor

Having established that the random walk behaviour exhibits deterministic chaos, we are interested to analyse the robot's phase space. First, we reconstruct the attractor, using the time-lag embedding method described in Section 4.2.3. The result is shown in Figure 5.6.

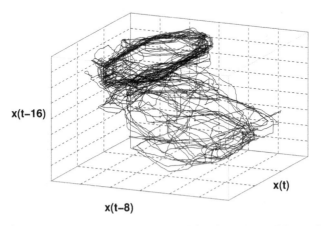

Figure 5.6. The three-dimensional reconstruction of the phase space of the random walk behaviour shows an attractor that essentially consists of two loops, one much more clearly defined than the other

We will now establish whether the robot's motion is periodic or not, by estimating the dimension of the attractor shown in Figure 5.6.

Dimension of the Attractor

We have by now established that the robot's random walk obstacle avoidance behaviour is deterministic (not really a surprise, because the robot's current position is dependent on the robot's previous positions, as the robot is physically unable to hop around randomly) and stationary. We have estimated that the prediction horizon is about 70 s, and that the Lyapunov exponent is about 0.07 bits/s. These are two quantitative descriptions of the attractor underlying the random walk obstacle avoidance behaviour, and we would now like to use a third, the correlation dimension.

The computation of the random walker's correlation dimension was already discussed in Section 4.5.1. We then estimated $2.7 < d < 3.3$, indicating that the behaviour of the random walker is aperiodic.

5.1.5 Random Walk: Summary of Results

In summary, we find that the random walk robot behaviour shown in Figure 5.1 is deterministic and stationary. We argued earlier that it would be beneficial if we had quantitative descriptions of robot-environment interaction, and we now have three: The prediction horizon of this behaviour is about 70 s, the Lypunov exponent about 0.07 bits/s, and the correlation dimension about 3.0.

5.2 "Chaos Walker"

Figure 5.7 shows the trajectory of the Magellan robot *Radix* that was programmed using the strategy shown in Table 5.1. Essentially, this strategy involves a straight move for a short period of time, and then a turn on the spot, where the turning direction and turning duration is determined by the chaotic quadratic iterator given in Equation 4.14.

The robot's position was logged every 250 ms. The robot's environment is shown in Figure 5.8.

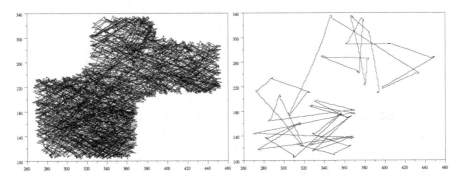

Figure 5.7. Trajectory of the "chaos walker". The entire trajectory is shown on the left, 5 min of robot motion on the right

Radix therefore moved predictably in a straight line for a little over 3 s, then turned on the spot for a time determined by the chaotic equation of the quadratic iterator given in Equation 4.14. We are interested to analyse this behaviour, and to describe it quantitatively. To do this, we will use the robot's motion along the x-axis, which is shown in Figure 5.9.

5.2.1 Stationarity

First, we will establish whether $x(t)$ is stationary or not. Dividing the entire time series $x(t)$ into three regions of equal size, we determine through a non-parametric analysis of variance (Section 3.4.4) that mean and standard deviation for the three sections are *not* the same, meaning that the signal is not stationary.

Figure 5.8. The environment in which *Radix* performed the chaos walker behaviour

Table 5.1. Pseudo code of the "Chaos Walker" behaviour

```
d(1)=0.125
t=1
while(1)
        t=t+1
        If obstacle detected
                Perform standard obstacle avoidance action
        else
                Move forward for 3.3 s
                d(t)=4d(t-1)(1-d(t-1))
                if d(t)>0.5
                        TurnTime=(PI * d(t) / 0.15) seconds
                else
                        TurnTime=-(PI * d(t) / 0.15) seconds
                Turn on spot at constant speed for TurnTime s
        end
end
```

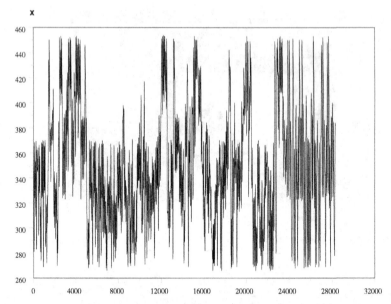

Figure 5.9. The x-coordinate of the motion shown in Figure 5.7

However, as discussed on page 100, the signal $\frac{dx}{dt} = \dot{x}$ is often stationary. This signal is shown in Figure 5.10, and indeed it turns out to be stationary, using the non-parametric analysis of variance. We will therefore analyse $\dot{x}$.

5.2.2 Determinism

To establish whether $\dot{x}$ (Figure 5.10) is deterministic or not, we'll apply the techniques discussed in Section 4.3.2, computing the return plot of $\dot{x}(t)$ vs $\dot{x}(t+2)$, and the ratio of ϵ/ϵ_b.

Both the return plot of $\dot{x}(t)$ vs $\dot{x}(t+2)$ and $\epsilon/\epsilon_b = 0.05$ indicate that $\dot{x}$ is deterministic, rather than stochastic (Figure 5.11 — compare with Figure 4.11).

5.2.3 Predictability of the "Chaos Walker"

Having established that our data is amenable to the mechanisms discussed in Chapter 4, we will now try to make some quantitative descriptions of the data. We'll start by looking at the predictability of $\dot{x}(t)$, as discussed in Section 4.4.2.

In order to determine the prediction horizon of $\dot{x}(t)$, we need to determine the correct embedding lag for reconstructing the attractor, which can be done by establishing when the autocorrelation reaches $1/e$ [Kaplan and Glass, 1995], or when the mutual information has its first minimum [Abarbanel, 1996]. The mutual information of $\dot{x}(t)$ is shown in Figure 5.12.

dx/dt

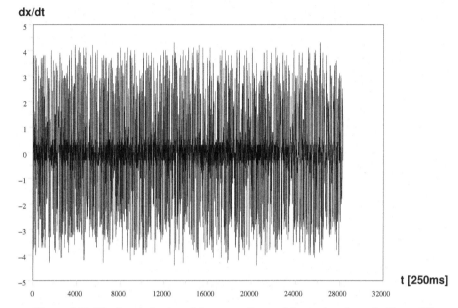

t [250ms]

Figure 5.10. The first derivative of the x-coordinate of the motion shown in Figure 5.7

$\dot{x}(t)$

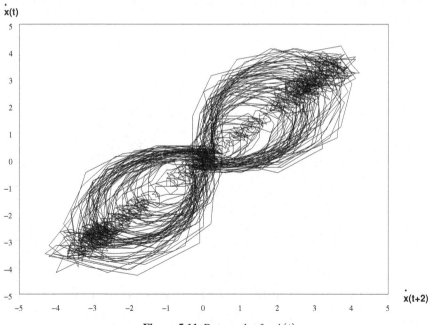

$\dot{x}(t+2)$

Figure 5.11. Return plot for $\dot{x}(t)$

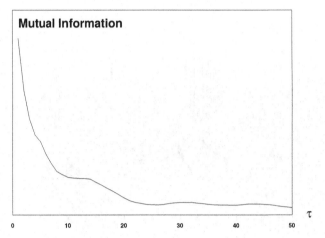

Figure 5.12. Mutual information of $\dot{x}(t)$ of the chaos walker

The mutual information has its first minimum for $\tau = 10$, Figure 5.13 shows a three-dimensional reconstruction of the attractor, using this embedding lag (the attractor actually has a dimension of around 4, so that Figure 5.13 is only one of many possible projections of the attractor onto 3D space).

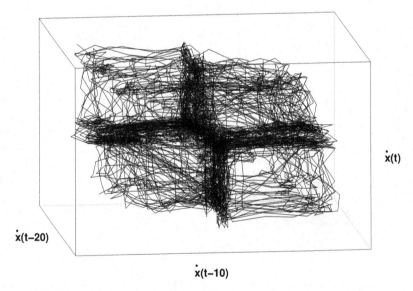

Figure 5.13. Reconstruction of the attractor of the chaos walker, underlying the motion along the $\dot{x}$ axis. $\tau = 10$

We now predict the second half of $\dot{x}(t)$, using the first half as our model, by constructing for each point $\dot{x}(t)$ a three-dimensional embedding $\mathbf{D} = [\dot{x}(t), \dot{x}(t-10), \dot{x}(t-20)]$, and finding the nearest neighbour $\mathbf{D_n}$ to each $\mathbf{D}$ within the first half of our data. The successors of $\mathbf{D_n}$ are then used to predict the successors to $\dot{x}(t)$.

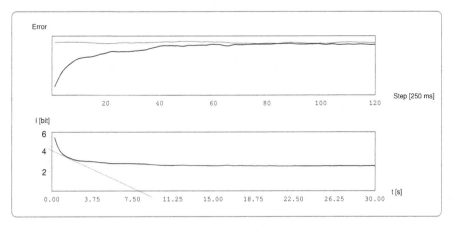

Figure 5.14. Prediction horizon for $\dot{x}$. The prediction horizon is about 19 s

The result is shown in Figure 5.14. The prediction horizon is about 19 s, and as there are about 6 bits of information available initially, we estimate the Lyapunov exponent from the prediction horizon as $\lambda \approx \frac{6}{19}$ bit/s = 0.3 bits/s.

The information loss shows a linear region (indicated in Figure 5.14), which has an information loss of $\lambda \approx \frac{4.3}{8}$ bit/s = 0.5 bit/s.

To get a third estimate, we use Wolf's algorithm to estimate λ, the results are shown in Table 5.2.3.

Using Wolf's algorithm we estimate $\lambda \approx 0.4$ bit/s, so that our final estimate of the Lyapunov exponent for the chaos walker is $0.3 < \lambda < 0.5$ bit/s.

5.2.4 Dimension of the Attractor Underlying the Chaos Walker

To conclude, we estimate the correlation dimension of the attractor shown in Figure 5.13, using Equation 4.23. Table 5.3 shows the result for various embedding dimensions p and correlation distances r. r_1 was selected such that it was about a quarter of the standard deviation of $\dot{x}$, r_2 was selected such that $C(r_1) \approx 5$ $C(r_2)$.

The results given in Table 5.3 show no very pronounced scaling region, and it is hard to give a precise dimension of the attractor, but based on the table we estimate that the dimension of $\dot{x}$ of the chaos walker is somewhere around 4.

Table 5.2. Estimates of $\lambda_{\dot{x}}$ of the chaos walker, using Wolf's algorithm. Embedding lag $\tau=10$, scalmn=0.1. Parameter settings that lie in a scaling region are printed in bold

p	evolv	scalmx	λ
3	3	0.8	1.9
	5		1.4
	7		1.2
	9		0.8
	11		0.7
	13		0.7
	15		0.7
	17		0.6
	19		0.5
3	13	0.3	1.0
		0.4	0.9
		0.5	0.8
		0.7	0.7
		0.9	0.7
		1.1	0.6
		1.3	0.6
		1.5	0.6
		1.7	0.6
		1.8	0.6
		2.2	0.5
3	13	1.4	0.6
4			0.5
5			0.4
6			0.4
7			0.4
8			0.4

Table 5.3. Correlation integral of $\dot{x}$ for various correlation distances r and correlation dimension (Equation 4.23) for $\dot{x}$ of the chaos walker

p	r_1	r_2	$C(r_1)$	$C(r_2)$	d
3	1.77	0.8	0.101	0.023	1.9
4	1.8	0.9	0.0443	0.0079	2.5
5	1.8	1.0	0.0179	0.00297	3.1
5	2.3	1.4	0.038	0.008	3.1
6	1.8	1.2	0.00695	0.00167	3.5
6	2.5	1.8	0.024	0.00695	3.8
7	1.8	1.2	0.0027	0.000587	3.8
7	2.5	1.8	0.011	0.0027	4.3
7	3.5	2.5	0.05	0.011	4.5
8	1.8	1.2	0.0011	0.00028	3.4
9	1.8	0.6	0.0005	0.00014	1.2

5.3 Analysing the Flight Paths of Carrier Pigeons

Carrier pigeons *Columba livia* f. *domestica* have an amazing ability to return home to their loft, when released at sites completely unknown to them, in some cases hundreds of kilometres away from home. To achieve this they use their innate magnetic compass, a learnt sun compass, and knowledge about navigational factors such as the distribution of the strength of the earth's magnetic field (for details   see [Wiltschko and Wiltschko, 2003]). Although most pigeons exhibit the homing ability reliably, there are differences between individual animals. It does matter if the release site is known to the pigeon, if the pigeon is experienced, what the weather is like, if the pigeon meets other pigeons along the way, *etc*. Some of these factors stem from the environment — they can't easily be taken into account when analysing pigeon behaviour — but others are specific to the individual, and it will be interesting to use the methods described in this book to highlight differences in the behaviour of individual animals.

In this case study, we will compare the homing behaviours of two individual pigeons. Our goal is to identify similarities and differences between the two homing behaviours; in other words: to characterise the pigeons' behaviour.

5.3.1 Experimental Procedure

To conduct this research, a miniature GPS logging device was attached to the pigeons. The animals were then taken to a release site approximately 21 km away from the loft, released, and during their flight home the pigeons' positions (longitude and latitude) were logged every second. These coordinates were subsequently converted into "earth-centred earth fixed" (ECEF) coordinates. The two flight paths we will analyse in this case are shown in Figure 5.15[1].

Both pigeons *569* and *97* were released the same distance from home, and both managed to fly home in a fairly straight path. Therefore, they seem to be very similar to each other — and yet, the following analysis will reveal important differences in their behaviour.

5.3.2 Analysis

Before we analyse the flight paths shown in Figure 5.15, a disclaimer: we will analyse the paths as they are, irrespective of the experimental conditions that prevailed when the data was obtained. We will assume that the *entire* paths contain "meaningful" information, characteristic of the behaviour of the individual pigeon. As it will turn out, both flight paths show two very distinct phases with

---

[1] I am grateful to Karen von Hünerbein and Roswitha and Wolfgang Wiltschko of the J.W. Goethe University of Frankfurt for making these data available to me.

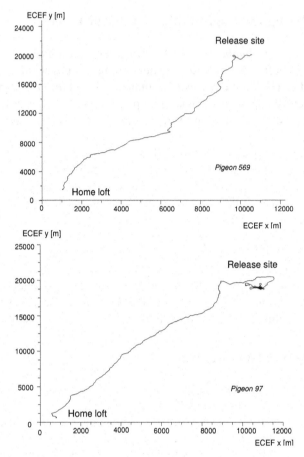

Figure 5.15. Homing flight paths of pigeon *569* (*top*) and pigeon *97* (*bottom*)

different characteristics — for the analysis here we assume that these two distinct phases are both relevant parts of the pigeon's behaviour, and not due to some experimental parameter, such as perhaps weather, light, noise, or anything like that.

For analysing the paths shown in Figure 5.15, the first question that needs to be addressed is "which datum contains relevant information about the pigeon's behaviour?" As we are looking at homing behaviour here, one meaningful datum is the deviation from the homeward direction at each point in time, *i.e.* the difference between the heading flown by the bird and the homeward direction. These deviations are shown in Figure 5.16, the dashed zero-degree line in the Figure denotes the direcction to home.[2]

---

[2] The data shown in Figure 5.16 has been median-filtered over a window of 19 s.

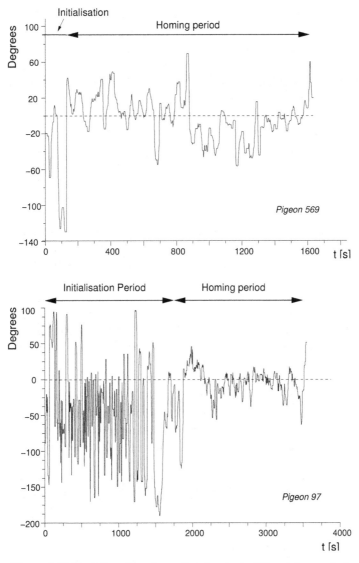

Figure 5.16. Deviations from home direction (note different time scales)

In order to apply the methods discussed in Chapter 4 to the time series shown in Figure 5.16, we first check that they are deterministic and stationary (see Section 4.3.2). It turns out that both deviations are deterministic ($\epsilon/\epsilon_b = 0.29$ and $\epsilon/\epsilon_b = 0.23$ for pigeon *569* and *97* resp.), but that only the deviation for pigeon *569* passes the runs test for stationarity (see Section 4.3.2). The deviation from home for pigeon *97* is stationary for the two individual sections (initialisation phase and homing phase), but not over the entire run. It is, however, weakly stationary, in that the mean over the entire run is roughly the same. The conclusion we draw from this is that the two birds differ in their behaviour, and that the results obtained for pigeon 97 have to be interpreted with caution.

Looking at Figure 5.16, a remarkable difference between pigeon 569 and pigeon 97 becomes clear: while initially both pigeons spend some time at the release site, pigeon 569 heads straight home after less than 200 s (i.e. its heading starts varying more or less symmetrically around the zero degree deviation line), whereas pigeon 97 spends 10 times as much time at the release site before it heads home. The reasons for this difference are unknown to us — it could be that pigeon 569 is the better navigator, or it could equally well be the case that pigeon 97 simply chose to spend more time with conspecifics at the release site before heading home. Whatever the reason, the different behaviour at the release site of the pigeons is clearly visible.

This difference becomes clear also if we analyse the dynamics of the pigeons' homing behaviour. The most obvious illustration of the difference is the phase space reconstruction of the data shown in Figure 5.16, which is given in Figure 5.17. Reconstructing the phase space through time lag embedding reveals the temporal relationship between deviations over time: in a pigeon that flies in an absolutely straight line home (zero degrees deviation from home direction throughout), the phase space should be a point of dimension zero, at position (0,0,0). The larger the variation of the pigeon's heading from true home, the "fuzzier" and higher-dimensional will the state space be. This is indeed visible in Figure 5.17.

The phase space of pigeon 569's deviation has a clearly defined, confined region around point (0,0,0) which corresponds to the pigeon's homing period. Pigeon 97's phase space, on the other hand, indicates that the deviation at time t is almost unrelated to the deviation at time $t - \tau$: the phase space is diffuse and of a larger dimension.

One quantitative descriptor of the phase spaces shown in Figure 5.17 is their correlation dimension, as introduced earlier in Section 4.5. Figure 5.18 shows the computation of these for the two homing behaviours, and indeed, while pigeon 569's attractor has a correlation dimension of approximately 2.3, the attractor of pigeon 97 has a correlation dimension of about 6.6!

Finally, we will look at the sensitivity to initial conditions, and the predictability of the birds' deviation. Figure 5.19 shows the prediction horizons for both homing behaviours.

Figure 5.19 shows that the homing behaviour of pigeon 569 is very predictable, with a prediction horizon of at least 40 s and a Lyapunov exponent of less than 0.1 bit/s, while the homing behaviour of pigeon 97 is far less predictable (prediction horizon about 15 s, Lyapunov exponent around 0.3 bit/s). Using Wolf's algorithm [Wolf, 2003], we confirm that for pigeon 569 $\lambda \approx 0.1$ bit/s, and for pigeon 97 $\lambda \approx 0.3$ bit/s.

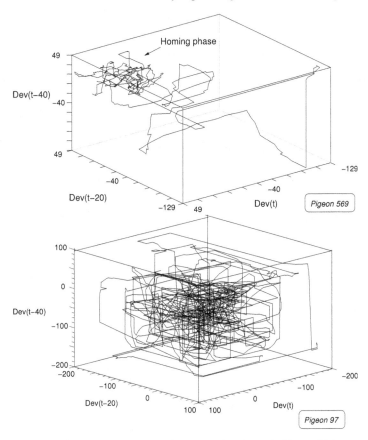

Figure 5.17. Phase space of the time series shown in Figure 5.16

5.3.3 Summary and Conclusion

In this last case study we compared the homing behaviour of two carrier pigeons. Specifically, we compared the two pigeons' deviation from the home direction *vs* time.

Although both pigeons arrive at the loft successfully, and although the two flight paths (Figure 5.15) look similar, we soon realise that the two homing behaviours differ considerably. Pigeon *569* heads straight home after less than 200 s at the release site, whereas pigeon *97* spends more than 1400 s before flying straight home. It is this initialisation period that differentiates the two pigeons; looking at their homing period alone, they look quite similar indeed.

Assuming that the initialisation phase is a descriptive part of the pigeons' behaviour, rather than an experimental fluke, we reconstruct and analyse the phase space of both pigeons' deviation from the true home direction, and find, not surprisingly, that the attractors describing the homing behaviours differ. Pigeon

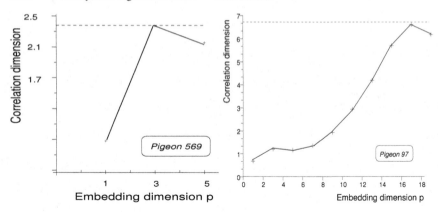

Figure 5.18. Correlation dimensions of the attractors shown in Figure 5.17 (note different scales)

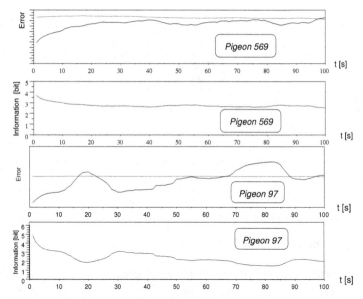

Figure 5.19. Prediction horizons of the attractors shown in Figure 5.17

569's attractor has a correlation dimension of around 2.3 and shows a clearly defined area corresponding to homing behaviour, whereas pigeon *97*'s attractor is diffuse, has a correlation dimension of about 6.6, and no pronounced homing region.

Furthermore, pigeon *569*'s homing behaviour is well predictable ($\lambda <$ 0.1 bit/s), while pigeon *97*'s homing behaviour is highly unpredictable ($\lambda \approx$ 0.3 bit/s) and shows characteristics of deterministic chaos.

6

Computer Modelling of Robot-Environment Interaction

Summary. This chapter discusses modelling robot-environment interaction in general (*i.e.* motivation, data logging, sampling, time series analysis), and how robot-environment interaction can be modelled, using system identification techniques such as artificial neural networks, ARMAX and NARMAX models.

6.1 Introduction

In this chapter we will investigate three different methods of modelling robot-environment interaction, that is, obtaining a mathematical description of the relevant parameters that generate a robot's motion. There are two main reasons for doing this. First, modelling will provide us with computer simulations of the robot's essential properties, which simplifies the development of robot control code. Second, a model will retain, in abstraction, the important aspects of the robot's operation, and thus provide a essential tool for the analysis of behaviour and scientific robotics.

6.1.1 Motivation

Faithful Simulation

To conduct experiments with mobile robots can be very time consuming, expensive, and difficult. Because of the complexity of robot-environment interaction, experiments have to be repeated many times to obtain statistically significant results. Robots, being mechanical and electronic machines, do not perform identically in every experiment. Their behaviour sometimes changes dramatically, as some parameter changes. Examples are specular reflections off smooth surfaces, or the influence of environmental parameters such as dust, changing motor characteristics as battery charge changes, *etc.*

Such hardware-related issues make simulation an attractive alternative. If it was possible to capture the essential components that govern robot-environment interaction in a mathematical model, predictions regarding the outcome of experiments could be made using a computer instead of a robot. This is faster, cheaper, and has the additional benefit that simulations can be repeated with precisely defined parameters. This enables the user to identify the influence of single parameters upon performance, something that cannot be done with real robots (because there are never two truly identical situations in the real world).

There are many advantages to simulation, apart from precise repeatability, speed, simplicity and low cost. Provided a faithful model can be found, simulation is a means of making predictions of systems that are too complex to analyse, or for which there is no data (yet) to perform a rigorous analysis (*e.g.* space exploration before the first man ever entered space). Simulation allows the controlled modification of parameters, and this modification in turn can lead to a better understanding of the model. Simulations can be used for teaching and training, stimulating interest (*e.g.* games). "What if" scenarios can be analysed using models, and simulation can give insights into how to best break up a complex system into subcomponents.

Models as Scientific Tools

One main purpose of scientific methods in robotics is to *understand* robot-environment interaction, to be able to identify the main contributors to a robot's behaviour, and to make predictions about the robot's operation.

Robot-environment interaction is a highly complex, often chaotic process that is so intricate that often it cannot easily be investigated on the real robot, in the real world. One motivation of robot modelling, therefore, is to obtain abstracted, simplified models of robot-environment interaction, which are more amenable to rigorous analysis.

This approach is not new. [Schöner and Kelso, 1988] demonstrate that it is possible to understand behavioural patterns, mostly in living beings, by means of concepts taken from stochastic nonlinear dynamics. In many cases, complex behaviour can be described by dynamics of a much lower dimension, resulting in a smaller number of degrees of freedom ("slaving principle"), which simplifies the analysis of the system. This is the motivation behind robot modelling, too: to obtain a simplified, transparent and analysable model of robot-environment interaction.

Fundamental Modelling Scenario

The fundamental modelling scenario is shown in Figure 6.1. An input vector $u(t)$ is associated with the output vector $y(t)$; the objective of the modelling process is to "identify" the relationship between $u(t)$ and $y(t)$ as a recurrence relation. The term "system identification" is therefore used for this process.

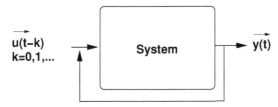

Figure 6.1. The fundamental simulation/computer modelling scenario

If the goal of the simulation is to obtain a faithful representation of the modelled system, then the model *has* to be constructed using *real* data, rather than general assumptions about robot and environment, due to the unpredictability of the real world, and the robot's susceptibility to noise and variation. This means, of course, that each model can only model one particular robot operating in one particular environment (see Figure 2.1)!

There are a number of possibilities to achieve this modelling task. A straightforward approach would be to log data at various locations in the real world, and at the prediction stage to use interpolation to predict the datum one is interested in. The difficulty in interpolation is that a lot of data has to be stored initially — the more closely spaced the "known" points are, the more precise the interpolation. Obviously, there is a limit to the amount of data that can be logged and stored, and therefore there is always a trade off between accuracy and efficiency.

There are (at least) three other possibilities for modelling the relationship between $\mathbf{u}(t)$ and $\mathbf{y}(t)$: using artificial neural networks (discussed below in Section 6.3), using linear polynomials (Section 6.4.2) and using non-linear polynomials (Section 6.5). After a discussion about data logging and sampling rates we we will discuss these options in turn.

6.2 Some Practical Considerations Regarding Robot Modelling

6.2.1 Data Logging Example: Determining the Sampling Rate

In this example we will determine a suitable sampling rate to analyse the behaviour of a wall-following robot. The robot's trajectory is shown in Figure 6.2; it is $x(t)$ that we want to analyse (right hand side of Figure 6.2).

In this case, data was sampled every 250 ms. The question is: is this sampling rate adequate for this experimental scenario, or is it too high or too low? If it is correct, then obviously $x(t)$ can be analysed straight away. If it is too high (oversampling), we can simply down sample the data by a rate of 1:n (use only ever nth data point). If we have under sampled, there is nothing but to repeat the experiment, and to log the robot's behaviour using a higher sampling rate.

Here are four ways of determining whether the sampling rate is suitable for the data or not:

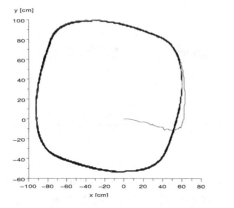
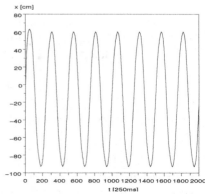

Figure 6.2. Movement of a wall-following robot, logged by an overhead camera (*left*), and the x-coordinate of that movement *vs* time (*right*)

1. Analyse the autocorrelation
2. Analyse the mutual information (see Equation 4.6)
3. Analyse the frequency spectrum
4. Use general physical considerations

Obviously, when logging data, it is essential to log only "meaningful" data points, *i.e.* every new data point should contain new, relevant information. If the sampling rate is too high, subsequent data points essentially contain the same information, and are therefore not "meaningful". The goal is to find the sampling rate at which subsequent data points contain novel information, and yet still capture the essence of the data.

The first two measures in the list above do this by computing the point at which the correlation between points $x(t)$ and $x(t + \tau)$ is near zero. Figure 6.3 shows that both autocorrelation and mutual information for $x(t)$ have fallen to near zero for $\tau \approx 65$. In other words, we now have an indication that "something interesting" happens no sooner than about every 16 s (65 steps at 4 Hz = 16.25 s).

Looking at the frequency spectrum of $x(t)$ (Figure 6.4), we can see that there are two relevant periods in our data, one repeating every 248 data points, and a smaller one repeating every 84 data points. These correspond broadly to the time taken by the robot to move from one corner of the arena to the next (84×0.25 s $= 21$ s), and to complete one round (248×0.25 s $= 62$ s).

In this case we get agreeing information from both the data and physical considerations. About every 20 s the robot turns a corner, and about every 60 s it therefore completes a round. Autocorrelation and mutual information both say that relevant new data points occur about every 16 s, broadly corresponding to the time elapsed between corners.

Usually an acceptable sampling rate in mobile robotics applications is about ten samples per period. The shortest period in this case is about 20 s. This means

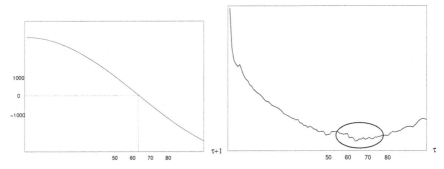

Figure 6.3. Autocorrelation (*left*) and mutual information of $x(t)$ shown in Figure 6.2. The autocorrelation falls to approx. zero and the mutual information has a minimum for a lag of $\tau \approx 65$

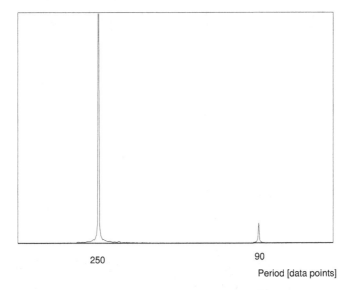

Figure 6.4. Frequency spectrum of $x(t)$

that we should log the robot's position about every 2 s — at a sampling rate of one sample every 250 ms this means we need to down sample at a rate of 1:8.

6.3 Case Study: Model Acquisition Using Artificial Neural Networks

This case study presents a mechanism that can learn location-perception mappings through exploration, and that is able to predict sensor readings for locations that have never been visited. A multilayer Perceptron is used to achieve this.

The structure of the two-layer network is shown in Figure 6.5. It is a multi-layer Perceptron that associates the robot's current position in (x, y) co-ordinates with the range reading of the one sonar sensor being modelled. Sixteen networks were used, one for each sonar sensor of a Nomad 200 mobile robot.

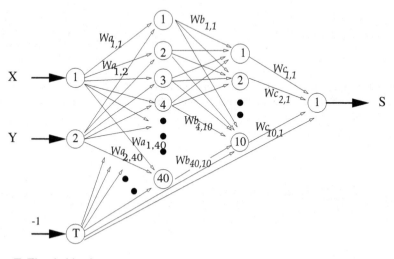

T: Threshold unit

W: Weight

S: Robot's sonar reading

X: Robot's x position in Cartesian system

Y: Robot's y position in Cartesian system

First hidden-layer units: 40

Second hidden-layer units: 10

Input units: 2

Output units: 1

Threshold units: 51

Threshold value: -1

Total weights: 490

Figure 6.5. The multilayer Perceptron used to acquire a model of the sonar sensor perception of a Nomad 200 robot

6.3.1 Experimental Procedure

To obtain training data, the robot was moved through the target environment in a methodical and regular manner, obtaining sonar sensor readings in regular intervals. The (x, y) location as obtained from the robot's odometry and the reading of the sensor were logged for later off-line training of the net. To minimise the error introduced by odometry drift, the robot's wheel encoders were frequently calibrated, and the path of the robot was chosen to be straight, rather than curved, which introduces less odometry error. Figure 6.6 shows two such experimental setups, indicating the paths the robot took for acquiring training data.

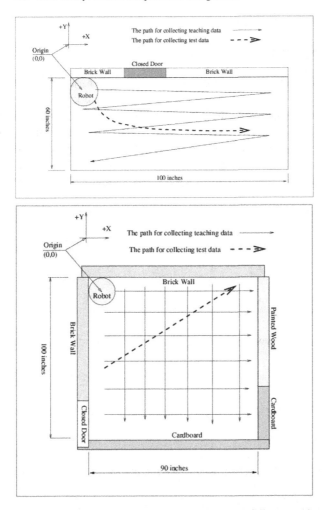

Figure 6.6. Experimental setup for data acquisition in two different environments

The robot was then led along a different, more "interesting" path to collect test data. As can be seen from Figure 6.6, training and test paths coincide in only very few points. If the acquired model has any *general* explanatory power about the robot's interaction with those two environments at all, it will be revealed when the network's predictions about sensory perception along the test path are compared with the actual sensory perception of the robot.

6.3.2 Experimental Results

Predicting Sensory Perception

Figure 6.7 shows those predictions *vs* the actual sensor readings for the test path given in Figure 6.6 (top). As can be seen, the network is able to predict the sudden increase in range reading at sample number 20, which is due to a specular reflection off the smooth wooden door. The (wrong) prediction made by a simplified numerical model ("Nomad simulator") is shown for comparison.

Likewise, Figure 6.8 shows the prediction of sensory perception of the learnt model along the test path given in Figure 6.6 (bottom).

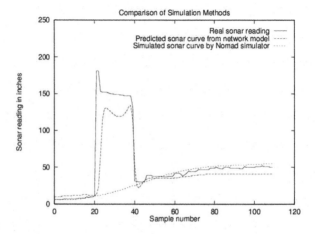

Figure 6.7. Simulation of sonar range readings during wall following. The sudden increase at sample number 20 ("Real sonar reading") is due to specular reflection off a wooden door. The generic simulator fails to predict this, because the model assumes uniform surface structure throughout the entire environment

Again, the acquired model is able to predict the sudden rise in range reading near location (400, -400).

Predicting Robot Behaviour

So far, there is an indication that the network model is able to predict the robot's sonar sensor readings in the target environment. Clearly this is useful, but really we are interested to predict the robot's *behaviour* in that environment, executing a particular control program.

For example, one could take a hardwired control program, *i.e.* a program that uses a fixed control structure without learning, to achieve a wall following

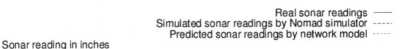

Sonar reading in inches

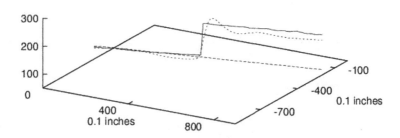

Figure 6.8. Network response to test data. A generic simulator is unable to predict the sudden increase in range reading due to specular reflection, while the learnt network simulator predicts this correctly

behaviour, and execute that program on the real robot, on its network simulation, and on the simple numerical simulator.

The result is astonishing! Because of the specular reflection off the door in the environment shown in Figure 6.6 (top), the robot actually crashes into the door, assuming there is more space ahead than there actually is. This is shown in Figure 6.9.

Because the simple numerical model assumes a uniform surface structure throughout the environment, it fails to predict that collision ("Nomad simulator" in Figure 6.9), whereas the network simulator sees it all coming. Figure 6.9 is one illustration of the fact that we are not necessarily talking about minor differences between the behaviour of a robot and its simulation: these are major discrepancies, leading to qualitatively completely different behaviour!

Let's run another simple hardwired control program, this time in the environment shown in Figure 6.6 (bottom). The program now is a "find-freest-space" program, which first takes all 16 sonar readings of the robot, then moves one inch in the direction of the largest reading, then repeats the process until either a distance of 100 inches has been covered, or the robot's infrared sensors detect an obstacle. This is a "critical" program, because even slight deviations will take the robot into a different area of the environment, resulting in a totally different trajectory. The results are shown in Figure 6.10.

In a uniform environment, one would expect that a find-freest-space program would lead the robot towards the geometrical centre of the environment, and

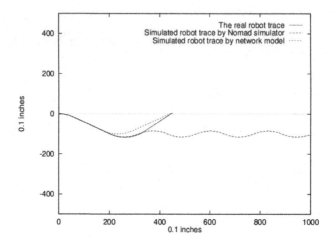

Figure 6.9. Because it assumes a homogeneous wall surface, a generic simulator fails to predict the collision that the real robot encounters, due to specular reflections. The learnt network simulator predicts the collision correctly

make the robot oscillate around that centre. This is precisely what the simple numerical simulator predicts.

However, in real life the robot actually moved towards the edge of the environment, which was predicted quite accurately by the network simulator.

Predicting the Behaviour of a Learning Controller

So far, the control programs we used to predict the robot's behaviour were relatively simple hardwired programs. These programs take sensor readings as inputs, and perform one specific, user-defined action in response.

Two components dominate the robot's behaviour in these experiments: the robot's sensory perception, and the control strategy used. Any simulation error will only affect the robot once when it uses a hardwired control program, *i.e.* in perception. The control program is user-supplied and fixed, and therefore not affected by simulation error.

If, on the other hand, we used a *learning* controller, any problems due to simulation errors would be exacerbated, in that first the robot would learn a control strategy based on erroneous sensory perceptions, and then it would execute that erroneous control strategy, taking erroneous sensor readings as input. Simulation errors have a double impact in these situations, and experiments with learning controllers could therefore serve very well as a sensitive measure of how faithful the simulator really is.

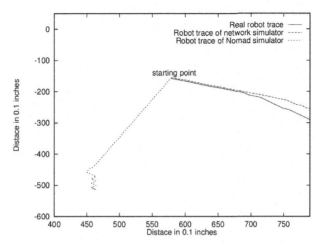

Figure 6.10. Simulated and real robot trajectories in response to the "find freest space" program. The simple numerical model predicts that the robot will move to the geometrical centre of the robot's environment, which is at (450, -450). The network simulator predicts a trajectory that is much closer to the trajectory actually taken

Experimental Setup

We therefore conducted experiments with an instinct-rule based learning controller similar to the one described in [Nehmzow, 2003a, p.76ff], using a Pattern Associator. The control strategy, therefore, was encoded in terms of the Pattern Associator's network weights. The objective of the learning process was to acquire a wall following behaviour.

Learning took place in simulation, either in the simple numerical model, or in the trained network model. The weights of the trained networks were then taken and loaded into the Pattern Associator of the real robot to control the robot's movements.

The trajectory of the real robot was then plotted against the trajectory that the respective simulators predicted. These trajectories are shown in Figure 6.11.

From Figure 6.11 one can see that the network simulator performs better than the simple numerical robot simulator. However, one can also see that our assumption of a more sensitive experiment due to the double impact of any error is true: in comparison with Figures 6.9 and 6.10 the predicted and simulated trajectory follow each other less closely here.

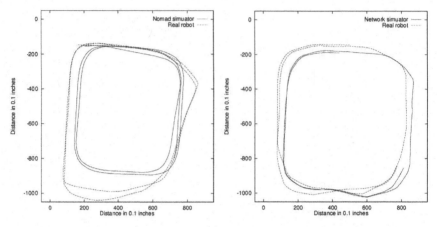

Figure 6.11. *Left:* The robot traces predicted by a generic simulator and the real robot using the weights generated by the generic simulator. *Right:* The network predicted robot trace is similar to the real one using the weights generated by the network simulator

6.4 Linear Polynomial Models and Linear Recurrence Relations

6.4.1 Introduction

Using real data to model the relationship between input variable $u(t)$ and output variable $y(t)$ has the advantage that the model, being based on real-world data obtained by the modelled agent operating in the target environment, is more faithful to the modelled behaviour than a generic model based on general assumptions. The case study given in the previous section demonstrates this.

However, modelling the opaque interdependency of robot, task and environment using an artificial neural network results in an opaque model. It is still not possible to *analyse* what really is going on — in order to do that, one would need an analysable mathematical model.

One possible way to obtain *analysable* models is to model the input-output relationship as a linear or nonlinear polynomial, known as ARMAX (autoregressive, moving average model with exogenous inputs) or NARMAX (nonlinear ARMAX).

There are considerable advantages to modelling input-output relationships using such transparent polynomial functions, rather than opaque mechanisms such as artificial neural networks:

1. The input-output representations are very compact and require very little space (memory) and processing time to compute.
2. They are amenable to rigorous mathematical analysis. For example, models of robot speed can be turned into models of robot acceleration by differentiating, models of acceleration can be modified to models of speed through

integration. Also, it is easier to estimate parameters such as Lyapunov exponent or correlation dimension from a closed mathematical function than from a time series.

3. Input-output relationships can be analysed graphically; plotting is straightforward, whereas in opaque models this is not possible.
4. The acquired model actually says something about the relationship between inputs and outputs. Parameters and lags indicate relevant process components. Questions like "What happens if a particular sensor fails?", "Which sensor is the most important (*i.e.* where would it be most effective to spend more money on a better sensor)?" or "What happens if the environment changes in a particular way?" can be addressed.
5. The analysis of similar behaviours, obtained by different means — for example achieving a particular robot behaviour through both a controller derived from control theory and one based on machine learning techniques — is easier when the models underlying those behaviours are considered: stability, sensitivity to noise, identification of relevant and irrelevant sensor signals are easier when a transparent mathematical expression is available for analysis.

The following sections will present two ways of obtaining such transparent models, ARMAX and NARMAX modelling. Both methods express the relationship between **u** and **y** as polynomials, ARMAX as a linear, and NARMAX as a non-linear polynomial. In both cases, models are transparent, and can be analysed systematically.

6.4.2 ARMAX Modelling

ARMAX (Auto-Regressive, Moving Average models with eXogeneous inputs) is a discrete time series model, commonly used in system identification, that models the relationship between the independent input variable $\mathbf{u}(t)$ and the dependent output variable $\mathbf{y}(t)$ as the linear polynomial given in Equation 6.1:

$$
\begin{aligned}
y_t = \quad & -a_1 y_{t-1} - a_2 y_{t-2} \ldots - a_i y_{t-i} \\
& +b_1 u_{t-1} + b_2 u_{t-2} \ldots + b_i u_{t-i} \\
& +d_1 e_{t-1} + d_2 e_{t-2} \ldots + d_i e_{t-i} + e_t
\end{aligned}
\tag{6.1}
$$

with **u** being the input, **y** being the output e being the noise model and a_k, b_k and d_k being the model parameters that have to be determined. This process is shown in Figure 6.12.

The ARMAX model has been widely studied for system identification, detailed information can be found for instance in [Pearson, 1999, Box et al., 1994].

The ARMAX model is limited, in that it is a linear model. However, as we will see, for many robotics modelling tasks a linear model is sufficient, and it is

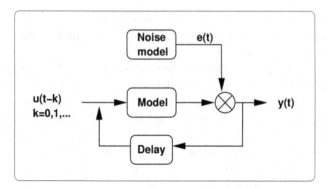

Figure 6.12. ARMAX system identification process

often possible to model input-output relationships such as sensory perception-motor response.

6.4.3 ARMAX Modelling Using Scientific Programming Packages

ARMAX system identification is part of many scientific programming packages (for example Scilab or Matlab), and can be used without any further programming. On page 176 there is an example of how this can be done by a one-line command in Scilab. The program listed below provides a slightly more user-friendly implementation of the ARMAX system identification process.

```
function [ypred, arc]=armaxid (y,u,r,s)
// (c) Ulrich Nehmzow
// ARMAX identification for n-dimensional input and one-dimensional output

// Program builds the model, using the first half of the time series
// then displays the entire time series against the model prediction

// u is the input signal of size   [(dimension n) (sample_size)]
// y is the output signal of size [(dimension 1) (sample_size)]
// r is the regression order on the output y,
// s the regression order on the input u

// ypred is the model-predicted output
// arc is the obtained ARMAX model

[a b]=size(y)
if (a>1)
        printf("Only one-dimensional outputs permitted! - Program aborted\n")
        abort
end

[ny,samples]=size(y)
[nu samples]=size(u)

// Check that both time series have even length - otherwise correct
if (samples/2-int(samples/2) ~= 0)
        y(samples+1)=y(samples)
        u(:,samples+1)=u(:,samples)
```

```
end

[ny,samples]=size(y)
[nu samples]=size(u)

modellength=int(samples/2)

// Add 2nd (empty) input line to make armax routine work
y=([y;zeros(1:samples)])

// Perform ARMAX identification, using the first half of the data
[arc,la,lb,sig,resid]=armax(r,s,y(:,1:modellength),u(:,(1:modellength)))
disp(arc)

// Now compute model-predicted values for the second half of the data
ypred=y(1,1:modellength)

for i=modellength+1:2*modellength
        ypred(i)=0
        // add the output-related components
        for lag=1:r
//              printf("Output component %f\n",la(1,2*lag+1))
                ypred(i)=ypred(i)-la(1,2*lag+1)*ypred(i-lag)
        end
        // add the input-related components
        for lag=0:s
//              printf("Lag %d\n",lag)
                for inp=1:nu
//                  printf("Input %d\n",inp)
                    ypred(i)=ypred(i)+lb(1,nu*lag+inp)*u(inp,i-lag)
//                  printf("Adding %f x %f\n",lb(1,nu*lag+inp),u(inp,i-lag))
                end
        end
end

// Now plot model-predicted output and actual output
xset("auto clear","off")
xbasc()
plot2d([modellength:samples],ypred(modellength:samples),3,
   rect=[modellength,min(min(y),min(ypred))-0.1,samples,
   max(max(y),max(ypred))+0.1])
plot2d([modellength:samples],y(1,(modellength:samples)),5,
   rect=[modellength,min(min(y),min(ypred))-0.1,samples,
   max(max(y),max(ypred))+0.1])
xtitle('','Data point','')
legends(['Original output' 'Model-predicted'],[5 3],3)
printf("Estimated standard deviation of noise and residual: %f\n",sig(1,1))
printf("Sum squared error: %f\n",sqrt(sum((ypred-y(1,:))^2)))

// Now compute errors if individual input components are reset to zero
ypred=y(1,1:modellength)

for block=1:nu
        for i=modellength+1:2*modellength
                ypred(i)=0
                // add the output-related components
                for lag=1:r
                        ypred(i)=ypred(i)-la(1,2*lag+1)*ypred(i-lag)
                end
                // input-related components
                for lag=0:s
                        for inp=1:nu
                            if (inp ~= block)
                                ypred(i)=ypred(i)+lb(1,nu*lag+inp)*u(inp,i-lag)
                            end
                        end
                end
```

```
          end
          printf("Blocking input %3d: error=%6.1f\n",block,norm(ypred−y(1,:),2))
          sse(block)=norm(ypred−y(1,:),2)
end

// Now print in format that is useful for publications

// Print input−related components
printf("b Matrix\n")
for lag=0:s
          printf("     t−%d   ",lag)
end
printf("SSE\n")
for inp=1:nu
          for lag=0:s
                  printf("%9.2f", lb(1,nu*lag+inp))
          end
          printf("%9.1f\n",sse(inp))
end

// Print output−related components
printf("\na Matrix\n")
for lag=1:r
          printf("  y(t−%d)   ",lag)
end
printf("\n")
for lag=1:r
          printf("%9.2f",la(1,2*lag+1))
end
printf("\n")
```

Example: ARMAX Modelling, Using Scientific Programming Packages

The following example demonstrates how linear polynomial models can be obtained, using Scilab.

Let's assume we have data that obeys the relationship given in Equation 6.2:

$$y(t) = 0.5u(t) - 0.3u(t-1) + 1.5u(t-2) - 0.7y(t-1) \qquad (6.2)$$

First, we generate a random input vector **u**:

```
u=rand(1:100);
```

We now compute the output variable y:

```
y(1)=0.2
y(2)=0.2
for i=3:100
        y(i)=0.5*u(i)−0.3*u(i−1)+1.5*u(i−2)−0.7*y(i−1);
end
```

The regression orders in Equation 6.2 are 1 for y and 2 for **u**. The following Scilab command will determine the ARMAX model that describes the data:

```
armax(1,2,[y';zeros(1:100)],u)
```

The result obtained is this:

```
A(z^-1)y=B(z^-1)u + D(z^-1) e(t)

A(x)  =

!   1 + 0.7x        1.329E-16x  !
!                               !
!   0               1          !


 B(x)  =

!                         2 !
!    0.5 - 0.3x + 1.5x     !
!                          !
!    0                     !
```

which is equivalent to Equation 6.3:

$$y(t) + 0.7y(t-1) = 0.5u(t) - 0.3u(t-1) + 1.5u(t-2) + noise, \quad (6.3)$$

from which follows $y(t) = 0.5u(t) - 0.3u(t-1) + 1.5u(t-2) - 0.7y(t-1) + noise$, which is the relationship expressed in Equation 6.2.

6.5 NARMAX Modelling

The ARMAX approach described above has the great advantage of determining *transparent* models of the y-**u** input output relationship, models that can be analysed and interpreted. However, its disadvantage is that ARMAX can only model linear input-output relationships. This is sufficient for some applications in robotics, but not all. This is where NARMAX (nonlinear ARMAX) comes in.

The NARMAX approach is a parameter estimation methodology for identifying both the important model terms and the parameters of unknown *nonlinear* dynamic systems. For single-input single-output systems this model takes the form of Equation 6.4:

$$\begin{aligned} y(k) = F[&y(k-1), y(k-2), ..., y(k-n_y), \\ &u(k-d), ..., u(k-d-n_\mu), \\ &e(k-1), \ldots, e(k-n_e)] + e(k), \end{aligned} \quad (6.4)$$

where $y(k)$, $\mathbf{u}(k)$, $e(k)$ are the sampled output, input and unobservable noise sequences respectively, n_y, n_u, n_e, are the orders, and d is a time delay. F[ ] is a nonlinear function and is typically taken to be a polynomial or a wavelet multi-resolution expansion of the arguments. Usually only the input and output measurements $u(k)$ and $y(k)$ are available and the investigator must process these signals to estimate a model of the system.

The NARMAX methodology breaks this problem into the following steps:

1. Structure detection
2. Parameter estimation
3. Model validation
4. Prediction
5. Analysis

These steps form an estimation toolkit that allows the user to build a concise mathematical description of the system [Chen and Billings, 1989]. The procedure begins by determining the structure or the important model terms, then continues to estimate the model parameters. These procedures are now well established and have been used in many modelling domains [Chen and Billings, 1989]. Once the structure of the model has been determined the unknown parameters in the model can be estimated. If correct parameter estimates are to be obtained the noise sequence e(k), which is almost always unobservable, must be estimated and accommodated within the model. Model validation methods are then applied to determine if the model is adequate. Once the model is accepted it can be used to predict the system output for different inputs and to study the characteristics of the system under investigation.

To discuss how to obtain a NARMAX model is beyond the scope of this book, but is widely discussed in the literature (see especially [Chen and Billings, 1989]).

We will now turn to applications of system identification techniques to robotics ("robot identification") . Particularly, we will demonstrate how robot identification can be used to simulate the operation of a mobile robot in a particular environment faithfully and accurately (environment identification), how robot identification can be used to facilitate cross-platform programming without actually writing robot code (task identification), and how it can be used to "translate" one sensor modality into another, allowing code that was written for a robot using one kind of sensor to be executed on a different robot that hasn't got this sensor (sensor identification).

6.6 Accurate Simulation: Environment Identification

6.6.1 Introduction

Our aim in environment identification is to derive accurate, transparent computer models of robot-environment interaction that can be used for code development: *generic* simulation programs are replaced by *specific* models of robot-environment interaction, derived from real-world data obtained in robotics experiments.

This section explains our procedure of deriving environment models, here using a simple robot behaviour in order to make the main mechanisms clear. Figure 6.13 shows the modelling relationship investigated in the experiments discussed in this first example.

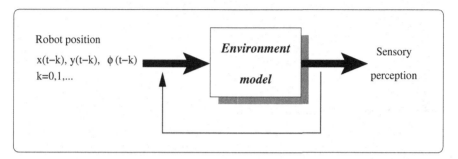

Figure 6.13. Environment modelling: a known function (such as the polynomial given in Table 6.1) maps robot position to sensory perception

In this example, we have chosen to investigate the wall-following behaviour of a Magellan Pro mobile robot (actually, the control code used to drive the robot was *not* a wall-following, but an obstacle-avoidance program; however, the interaction of our robot with its environment resulted in a wall-following trajectory). The robot used was the Magellan Pro shown in Figure 1.1, the trajectory we logged every 250 ms with an overhead camera is shown in Figure 6.14.

First, this data was subsequently subsampled at a rate of 1:15, so that the time elapsed between data points was 3.75s. The average speed of the robot in this experiment was 8 cm/s, so that the distance travelled between logged robot positions was about 30 cm.

To obtain the non-linear model, a NARMAX model identification methodology was followed. First, the model structure was determined by choosing regression order and degree of the inputs and output. "Degree" is defined as the sum of all exponents in a term, where a "term" is a mathematical expression as shown in each line of, for example, Table 6.1.

To determine a suitable model structure, we use the orthogonal parameter estimation algorithm described in [Korenberg et al., 1988]. This indicates (prior to the calculation of the model) which model terms are significant for the calculation of the output.

We then obtain the model, using the first half of the available data ("training data"), and validate it using the remaining half ("validation data").

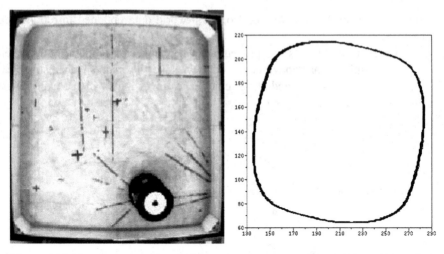

Figure 6.14. The environment in which experiments were conducted (*left*), and the robot's trajectory (*right*). The robot is visible in the *bottom right hand corner* of the left image

The resulting model is shown in Table 6.1, it computes the distance measured by the laser sensor at 67° from the left of the robot (L67, see Figure 6.15) as a function of its position (x, y).

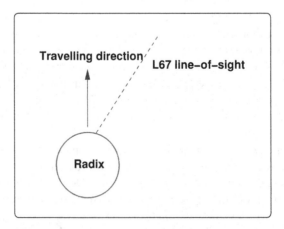

Figure 6.15. L67 (modelled in Table 6.1) is the robot's single-ray laser perception towards the right hand side of the robot

The comparison between the true laser perception and that predicted by the model of Table 6.1 is shown in Figure 6.16; it shows clearly that the robot's laser perception L67(t) can indeed be modelled as a function of the robot's (x, y)

Table 6.1. Parameters of a polynomial modelling the robot's single-ray laser perception L67 as a function of the robot's position (x,y). The time series of this model is shown in Figure 6.16. See also Figure 6.15

$$
\begin{aligned}
L67(t)= &+1.8801351 \\
&+0.0087641 * x(t) \\
&-0.0116923 * x(t\text{-}1) \\
&-0.0060061 * x(t\text{-}2) \\
&+0.0116420 * y(t) \\
&+0.0143721 * y(t\text{-}1) \\
&-0.0064808 * y(t\text{-}2) \\
&+0.0004983 * x(t)^2 \\
&+0.0021232 * x(t\text{-}1)^2 \\
&+0.0006722 * x(t\text{-}2)^2 \\
&-0.0002464 * y(t)^2 \\
&+0.0018295 * y(t\text{-}1)^2 \\
&+0.0015442 * y(t\text{-}2)^2 \\
&-0.0028887 * x(t) * x(t\text{-}1) \\
&+0.0023524 * x(t) * x(t\text{-}2) \\
&+0.0002199 * x(t) * y(t) \\
&-0.0025234 * x(t) * y(t\text{-}1) \\
&+0.0022859 * x(t) * y(t\text{-}2) \\
&-0.0029213 * x(t\text{-}1) * x(t\text{-}2) \\
&+0.0006455 * x(t\text{-}1) * y(t) \\
&+0.0014447 * x(t\text{-}1) * y(t\text{-}1) \\
&-0.0027139 * x(t\text{-}1) * y(t\text{-}2) \\
&-0.0004945 * x(t\text{-}2) * y(t) \\
&+0.0003262 * x(t\text{-}2) * y(t\text{-}1) \\
&+0.0009349 * x(t\text{-}2) * y(t\text{-}2) \\
&-0.0010366 * y(t) * y(t\text{-}1) \\
&+0.0013326 * y(t) * y(t\text{-}2) \\
&-0.0037855 * y(t\text{-}1) * y(t\text{-}2)
\end{aligned}
$$

position at times t and $t-1$, easily and accurately. Pearson's correlation coefficient between modelled and true data is 0.75 (significant, $p<0.01$).

It is interesting to note that the robot's orientation ϕ is not needed in order to model the perception of the laser sensor. The reason for this is the restricted motion of the robot (following the perimeter of the environment), which by specifying (x, y) essentially also specifies orientation ϕ, so that ϕ is not needed explicitly in the model.

6.6.2 Environment Modelling: ARMAX Example

For simple cases, the environment model does not have to be nonlinear, and in the following we will develop a linear ARMAX model of the laser perception a Magellan Pro mobile robot perceives as it follows a trajectory very similar to that shown in Figure 6.14.

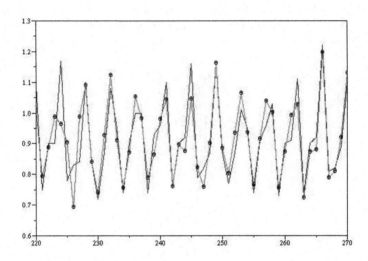

Figure 6.16. Modelling the robot's laser perception $L67$ as a function of position (see also Table 6.1). True sensor perception is shown as a line *with circles*, the model-predicted output as a line *without circles*

As the robot followed a trajectory like the one shown in Figure 6.14, the range data obtained from the laser range finder was logged every 250 ms, as was the robot's position $[x, y]$ and rotational velocity $\dot{\phi}$. This data was subsequently subsampled at a rate of 1:15 (see Section 6.2.1 for a discussion of sub sampling), and the laser perception of all laser values between "45° to the left" to "straight ahead" were averaged to obtain the laser perception $LM(t)$ shown in Figure 6.17. The robot's position $[x, y]$ and rotational velocity $\dot{\phi}$ are also shown in that figure.

We will now obtain an ARMAX model $LM(t) = f(x(t), y(t), \dot{\phi}(t), x(t - 1), y(t - 1), \dot{\phi}(t - 1))$, using the Scilab ARMAX package. We will use the first 500 data points to construct the model ('model data'), and the remaining 451 data points to validate the model ('validation data'):

```
armax(0,1,[y(1:500),zeros(1:500)']',
          [u(1:500,1),u(1:500,2),u(1:500,3)]')
ans  =  A(z^-1)y=B(z^-1)u + D(z^-1) e(t)

 A(x) =
 !  1      0 !
 !  0      1 !

 B(x) =
   0.0078763 - 0.0041865x
 - 0.0078344 + 0.0101931x
```

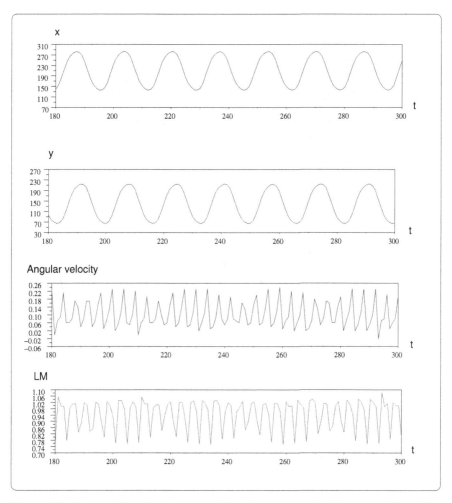

Figure 6.17. Robot position (*top three graphs*) and laser perception (*bottom graph*) used for environment identification

```
 - 1.3727936 - 0.1602578x

 D(x)    =
 !   1     0  !
 !   0     1  !

  e(t)=Sig*w(t); w(t) 2-dim white noise

            | 0.0558112  0 |
  Sig=  | 0           0 |
```

This results in the model given in Equation 6.5:

$$LM(t) = 0.0078763x(t) - 0.0041865x(t-1) \qquad (6.5)$$
$$-0.0078344y(t) + 0.0101931y(t-1)$$
$$-1.3727936\dot{\phi}(t) - 0.1602578\dot{\phi}(t-1)$$

Using this model to predict the laser perception of our validation data (data points 501 to 941), we obtain the result shown in Figure 6.18.

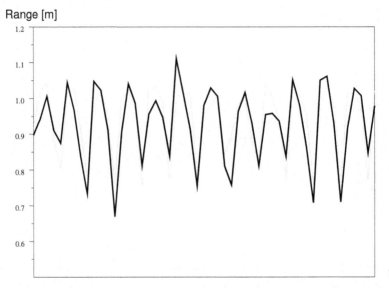

Range [m]

Figure 6.18. Actual laser perception (*thin, faint line*) *vs* model-predicted perception (Equation 6.5, *thick line*)

6.6.3 Localisation Through Environment Modelling

The modelling scenario shown in Figure 6.13 can be reversed, and used for self-localisation. If the relationship between the robot's location and the sensory perception obtained at that location is known, then the reverse is also true (provided the relationship is expressed in an invertible function), and the robot's position can be determined from sensory perception. This is shown in Figure 6.19.

Model 1

We will illustrate this process through an example.

Experiments were conducted in an enclosed arena of roughly 2×2 metres size, in which the robot performed a wall-following action. The motion of the

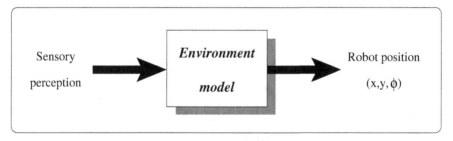

Figure 6.19. Perception-based localisation

robot, as well as all its sensory perceptions were logged every 6.25 s, using an overhead camera. Figure 6.20 shows a bird's eye view of the arena, the robot is also visible in that figure.

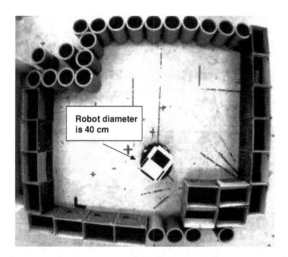

Figure 6.20. The experimental arena in which data for the localisation experiment was logged

The trajectory taken — ground truth obtained through the overhead camera — is shown in Figure 6.21.

The objective of this experiment is to establish the robot's position (x, y), using information from the 16 sonar sensors and a selection of 12 laser readings of the robot. We then use an ARMAX modelling process (shown in Figure 6.22) to obtain a model of position (x, y), given the raw laser and sonar perceptions indicated in Figure 6.22.

Using the program shown in Section 6.4.3, we obtain the models for $x(t)$ and $y(t)$ shown in Tables 6.2 and 6.3 respectively.

Even though the model requires a relatively high regression order, it *is* possible for the robot to self-localise in our experimental arena, using sensory infor-

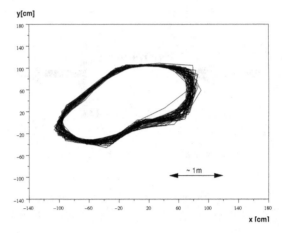

Figure 6.21. Trajectory of a wall following robot

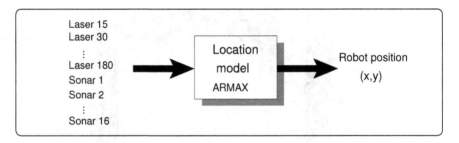

Figure 6.22. ARMAX identification task for self-localisation

mation alone. Figure 6.23 shows true $< x, y >$ vs $< \tilde{x}, \tilde{y} >$ predicted by the model. The Spearman rank correlation coefficients r_x between $x(t)$ and $\tilde{x}(t)$ and r_y between $y(t)$ and $\tilde{y}(t)$ are 0.96 and 0.95 respectively (significant, $p < 0.05$).

Figure 6.24 shows a comparison between the actual trajectory taken by the robot, and the trajectory predicted by the robot (using the test data). The mean localisation error is $29.5cm \pm 0.84cm$[1]. The distribution of localisation errors is given in Figure 6.25.

Model 2: Refining Model 1

One purpose of acquiring model 1 of $< x(t), y(t) >$, using most of the sensory information available to the robot, was to investigate whether all of this information is actually needed to localise. We can use Tables 6.2 and 6.3 to establish that it is not: in these tables the rightmost column indicates the sum-squared-error obtained when a particular model term is removed. The six sensor signals that

---

[1] For comparison, the robot's diameter is 40 cm.

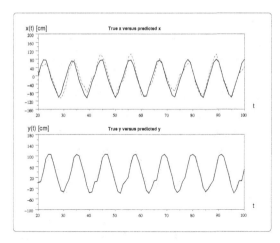

Figure 6.23. Actual robot position $< x(t), y(t) >$ (*thick, bold line*) *vs* the position that is estimated from sensory perception, using model 1 given in Tables 6.2 and 6.3 (*faint line*)

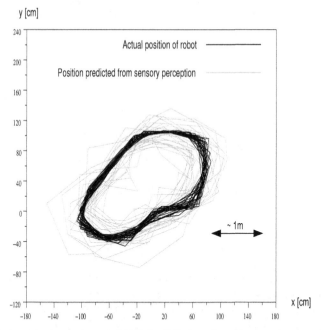

Figure 6.24. Actual robot trajectory (*thick, bold line*) *vs* the trajectory that is estimated from sonar and laser perceptions, using model 1 given in Tables 6.2 and 6.3 (*faint line*)

Table 6.2. ARMAX model 1 of $x(t)$ (*i.e.* $x(t) = -47.40 Laser15(t) - 89.78 Laser15(t - 1) \ldots$). SSE indicates the sum squared error if the respective term is removed from the model

	t	t-1	t-2	t-3	SSE
Laser 15	-47.40	-89.78	-72.11	-21.42	**2506.4**
Laser 30	-12.04	-14.14	-7.05	12.05	410.0
Laser 45	-17.19	-9.57	-10.05	-2.25	634.9
Laser 60	-3.37	4.99	-2.78	7.01	337.9
Laser 75	18.90	9.85	45.78	16.14	**1784.3**
Laser90	2.39	9.28	-7.01	-0.52	337.9
Laser 105	-14.71	-17.80	-9.74	-12.08	**1435.6**
Laser 120	23.16	19.02	18.91	-1.52	**1734.7**
Laser 135	-18.79	1.65	6.90	19.91	555.0
Laser 150	2.05	-6.87	-11.79	10.39	402.0
Laser 165	13.44	9.04	14.84	-2.08	**1195.2**
Laser 180	18.36	14.95	19.65	-5.86	**1597.4**
Sonar 1	-4.17	-3.14	-2.45	0.76	444.1
Sonar 2	-3.02	-1.93	-0.91	-1.19	474.0
Sonar 3	-3.27	1.52	2.98	2.15	364.0
Sonar 4	-0.92	-3.96	-4.57	-3.42	636.5
Sonar 5	-4.93	-5.51	-3.06	-1.11	710.4
Sonar 6	1.46	-0.80	-2.16	-3.92	418.0
Sonar 7	-1.02	-2.63	-4.06	-3.44	622.5
Sonar 8	0.99	0.99	0.99	0.99	428.8
Sonar 9	1.24	0.50	1.16	0.48	346.5
Sonar 10	1.15	2.72	0.59	-0.07	356.7
Sonar 11	-0.17	-0.39	-0.55	-0.41	317.5
Sonar 12	-0.39	0.63	0.46	1.39	322.5
Sonar 13	-0.49	0.94	2.83	-0.07	331.3
Sonar 14	1.97	2.20	2.78	1.79	374.3
Sonar 15	0.44	2.47	1.16	0.71	357.2
Sonar 16	-3.86	-1.90	-1.01	-0.88	422.4

have the highest contribution to the model account for the majority of all contributions to the model. Interestingly, sonar signals come nowhere near the top; they obviously contain too much imprecise or contradictory information to be as useful in the localisation process as the laser perceptions.

Based on this consideration, we obtained a refinement of model 1 by only using the six most important sensor signals. This second model is shown in Tables 6.4 and 6.5. From these tables we see that less sensor information is needed to localise, but that much higher regression orders are now required: the robot is using less sensor information, but over a longer period of time. A time window of 13 samples equates to over 80 s of movement!

Figure 6.26 shows actual and predicted positions; as in model 1 the correlation is highly significant.

Table 6.3. ARMAX model 1 of $y(t)$

	t	t-1	t-2	t-3	t-4	*sse*
Laser 15	35.65	7.86	-38.68	-74.11	-38.72	**1223.4**
Laser 30	8.65	-5.01	-7.89	-12.35	-3.82	378.9
Laser 45	8.18	-9.80	-0.75	-3.61	1.90	295.0
Laser 60	-5.96	-0.57	13.33	10.59	-6.89	317.6
Laser 75	3.74	5.79	0.27	29.41	26.38	**1285.7**
Laser 90	-11.34	1.64	1.91	-1.28	-1.10	358.7
Laser 105	3.52	-3.97	-14.41	-8.67	-5.30	817.6
Laser 120	7.41	17.19	20.37	16.40	13.45	**2114.1**
Laser 135	4.13	-7.20	-12.44	-9.44	9.32	623.9
Laser 150	0.58	10.29	-2.45	-8.90	1.23	320.3
Laser 165	-6.16	1.86	11.38	15.51	0.53	794.1
Laser 180	-10.22	2.12	4.88	18.30	9.67	852.5
Sonar 1	-1.64	-2.72	-1.97	-2.74	-1.15	455.6
Sonar 2	-0.42	-1.85	1.14	1.18	-0.66	280.5
Sonar 3	-1.98	-3.82	-0.68	0.99	0.72	353.7
Sonar 4	2.50	1.30	0.37	-3.46	-4.25	349.6
Sonar 5	0.16	-1.91	-2.42	-1.83	-1.61	446.0
Sonar 6	2.67	1.77	0.41	-0.79	-1.54	307.3
Sonar 7	2.12	1.56	-1.13	-2.55	-1.00	297.4
Sonar 8	-2.93	-2.93	-2.93	-2.93	-2.93	**1069.0**
Sonar 9	0.75	0.35	1.02	1.72	1.56	333.0
Sonar 10	-1.07	-0.35	2.37	1.30	1.37	301.0
Sonar 11	0.24	1.19	0.66	-0.86	0.73	282.2
Sonar 12	-0.06	0.48	0.61	0.61	1.72	285.4
Sonar 13	-1.29	0.14	1.48	3.80	2.89	318.9
Sonar 14	-1.95	0.19	1.61	2.51	2.01	292.1
Sonar 15	-0.77	0.15	0.96	1.11	0.58	284.3
Sonar 16	-1.21	-2.46	-1.59	-1.48	-0.09	392.6

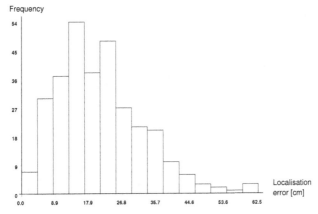

Figure 6.25. Distribution of localisation errors for the test data, using model 1 given in Tables 6.2 and 6.3

Table 6.4. Alternative model (model 2) to determine x from laser sensor signals

	t-0	t-1	t-2	t-3	t-4	t-5	t-6	t-7	t-8	t-9	t-10	t-11	t-12	SSE
Laser 15	-55.94	-73.11	-75.63	-53.57	-5.18	44.67	67.42	68.19	50.96	26.88	-13.36	-39.01	-58.16	1541.2
Laser 75	-2.26	3.21	7.64	4.45	-2.06	-6.28	-12.83	-13.37	-12.57	-12.72	4.14	12.57	16.95	388.3
Laser 105	-13.16	-12.03	-8.71	-4.58	6.79	5.94	8.57	5.79	-0.24	-5.02	-13.64	-13.53	-8.21	1388.3
Laser 120	18.13	13.45	7.03	-7.22	-13.13	-18.91	-16.63	-6.17	5.74	10.95	8.52	6.19	5.97	510.7
Laser 165	11.64	15.93	13.81	9.17	5.46	0.70	-9.99	-7.16	-7.12	2.17	12.50	9.04	9.41	2133.5
Laser 180	-3.53	-0.52	-2.58	-2.42	0.12	3.72	-1.11	-5.95	-3.73	-0.24	1.46	7.79	12.13	354.3

Table 6.5. Alternative model (model 2) to determine y from laser sensor signals

	t-0	t-1	t-2	t-3	t-4	t-5	t-6	t-7	t-8	t-9	t-10	t-11	t-12	t-13	SSE
L. 15	12.36	-13.84	-39.92	-71.09	-64.45	-45.60	7.85	37.98	47.64	61.94	48.53	12.55	-11.17	-41.20	961.4
L. 75	0.51	0.05	-0.13	13.07	8.17	4.18	2.70	-9.25	-9.91	-11.62	-15.66	-5.29	6.28	10.47	274.3
L. 105	6.92	-5.11	-9.54	-7.95	-8.41	-1.59	-0.17	8.40	8.97	8.41	3.14	-9.05	-11.26	-10.78	784.1
L. 120	6.13	18.91	15.14	10.25	4.96	-0.88	-15.79	-14.63	-12.42	-5.81	4.24	8.46	6.59	6.44	926.9
L. 165	-8.35	3.67	10.91	14.84	11.12	6.60	4.76	-6.42	-6.92	-10.99	-3.88	7.06	7.16	8.07	1245.9
L. 180	-4.24	-1.22	-1.57	-0.83	-0.78	0.06	0.61	-0.27	-2.01	-2.64	0.20	-2.38	3.65	10.78	263.8

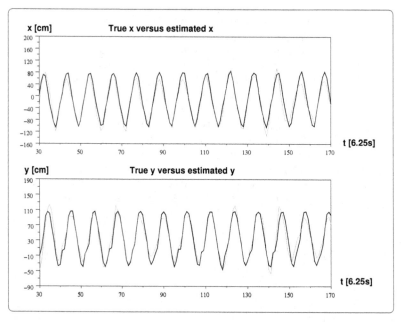

Figure 6.26. Actual robot position $< x(t), y(t) >$ (*thick, bold line*) vs the position that is estimated from sensory perception, using the alternative model 2 given in Tables 6.4 and 6.5 (*faint line*)

Figure 6.27 shows the actual trajectory taken vs the trajectory predicted by the model. The mean localisation error is lower than in model 1: 22 cm $\pm$ 0.7 cm. The distribution of errors is shown in Figure 6.28.

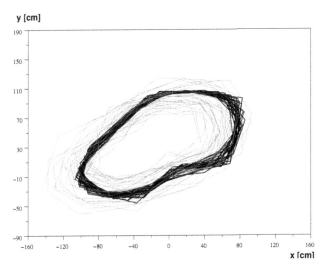

Figure 6.27. Actual robot trajectory (*thick, bold line*) *vs* the trajectory that is estimated from laser perception, using model 2 given in Tables 6.4 and 6.5 (*faint line*)

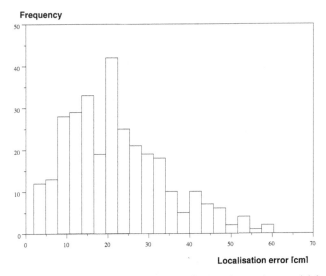

Figure 6.28. Distribution of localisation errors for the test data, using model 2 given in Tables 6.4 and 6.5

Model 3: Refining Model 2 Further by Taking Motion Into Account

Models 1 and 2 demonstrate that a localisation accuracy of about half the robot's diameter can be achieved by either observing most of the robot's sensors over a time window of about 24 s, or observing the most informative sensors over a period of about 80 s. However, wall following behaviour is clearly highly repeatable and predictable, and if regression on the model output is included — making past predictions of x and y part of the model — the model ought to shrink and become more precise. To investigate this hypothesis was the purpose of the third experiment.

Tables 6.6 and 6.7 show that indeed we can now build a model which still only uses the six most important sensor signals, as well as past predictions of x or y respectively, but that now a regression order of 4 (24 s) is sufficient.

Table 6.6. Model 3 for $x(t)$, taking previous x estimates into account

	t	t-1	t-2	t-3	t-4
x		1.06	- 0.72	0.65	- 0.70
L15	- 28.92	- 7.01	- 0.54	- 6.80	
L75	- 2.99	+ 7.08	- 2.78	+ 10.33	
L105	3.91	- 1.50	- 3.49	+ 1.68	
L120	3.63	+ 1.28	- 3.85	- 0.24	
L135	- 7.40	+ 2.14	+ 6.34	+ 0.01	
L180	2.59	+ 3.32	- 8.06	+ 3.42	

Table 6.7. Model 3 for $y(t)$, taking previous y estimates into account

	t	t-1	t-2	t-3	t-4
y		0.85	- 0.18	0.09	- 0.49
L15	9.96	- 5.83	- 15.84	- 8.31	+ 0.88
L30	16.80	- 9.71	- 5.20	+ 2.48	+ 0.34
L75	1.46	+ 0.41	- 5.68	+ 7.11	- 0.48
L120	1.95	+ 2.97	+ 4.06	- 4.00	+ 7.88
L135	1.99	+ 1.03	- 5.28	+ 2.58	+ 5.50
L180	- 2.62	+ 1.80	+ 4.10	- 3.82	+ 0.34

Figure 6.29 shows actual and predicted trajectory in this case. The localisation error has now almost halved to 13 cm $\pm$ 0.4 cm, and the correlation coefficients r_x and r_y are both 0.98 (significant, $p < 0.05$). The distribution of localisation errors is shown in Figure 6.30.

Model 4: Refining Model 3 Further

The refinement of model 3 is almost a formality, but yields considerable improvements again. Using not only past values of the predicted variable, but also the robot's position $<x, y>$, it is possible to obtain a very compact model that allows localisation to very high accuracy.

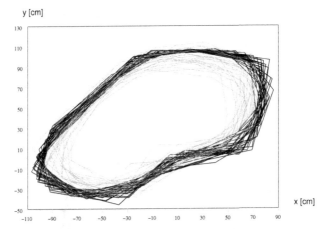

Figure 6.29. Actual robot trajectory (*thick, bold line*) *vs* the trajectory that is estimated from sonar perception, using model 3 given in Tables 6.6 and 6.7 (*faint line*)

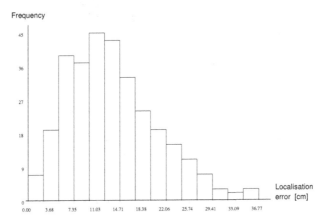

Figure 6.30. Distribution of localisation errors for the test data, using model 3 given in Tables 6.6 and 6.7

Equations 6.6 and 6.7 show the result. Eight inputs over a time window of 6 s are now sufficient to establish the robot's location with a mean localisation error of 10.5 cm $\pm$ 0.4 cm.

$$x(t) = 1.04x(t-1) - 0.65y(t-1) \tag{6.6}$$
$$-6.7L_{15}(t) - 8.84L_{30}(t) + 4.05_{L75}(t)$$
$$+7.1L_{120}(t) - 5.75L_{135}(t) + 14.53L_{180}(t)$$

$$y(t) = -0.49x(t-1) - 0.56y(t-1) \tag{6.7}$$

$$-2.54L_{15}(t) + 11.43L_{30}(t) + 2.19L_{75}(t)$$
$$+5.93L_{120}(t) - 2.14L_{135}(t) + 3.10L_{180}(t)$$

The predicted and actual trajectories taken are shown in Figure 6.31, in contrast to localisation based on perception alone, the two trajectories now resemble each other very closely indeed.

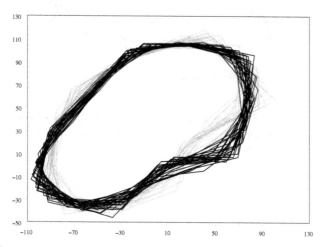

Figure 6.31. Actual robot trajectory (*thick, bold line*) *vs* the trajectory that is estimated from previous position $< x, y >$ and laser perception, using model 4 given in Equations 6.6 and 6.7. The *solid line* denotes the actual trajectory, the *faint line* the predicted trajectory. The mean localisation error is 10.5 cm $\pm$ 0.4 cm.

The distribution of localisation errors for model 4 is shown in Figure 6.32.

Conclusions

As we have seen in this example, sensor-based localisation is possible, using AR-MAX models. The section discussed four models of decreasing complexity, but increasing precision, demonstrating which components of the robot's perceptual vector are useful for self-localisation, and which are not.

In the first model we demonstrated that sensor-based localisation *is* possible in the arena we used. Model 1 also revealed which sensor information is particularly useful for that task: the six most useful senses are the laser perceptions facing the nearside wall. Sonar perceptions turn out to be not particularly informative regarding position.

Based on these considerations we developed a second model that uses an impoverished perception input. In this model, we find that higher regression orders are necessary to retain the same localisation accuracy. In other words: using sensor information alone, the roboticist has the choice of either using all sensor

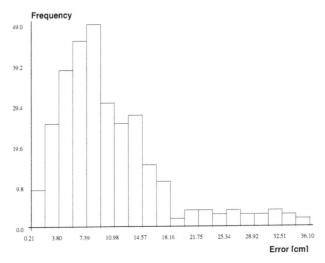

Figure 6.32. Distribution of localisation errors for the test data, using model 4 given in Equations 6.6 and 6.7

information available over a shorter time window (24 s), or using less sensor information over a longer time window (80 s).

Models 3 and 4 demonstrate that in the case of wall following, very accurate and compact models can be obtained by regressing not only over sensor perceptions, but also over past position estimates. The reason for this observation is, obviously, that wall following is a highly repetitive and predictable behaviour. The final model achieves a mean localisation accuracy of $\frac{1}{4}$ of the robot's diameter, using a model that contains only 8 terms and a regression of only 6 s.

6.7 Task Identification

In task identification the objective is to obtain a model of the control program of the robot. This results in the "compression" of program code into a single polynomial equation. An immediate advantage in doing this is the ease of communication of a robot task in cases where the actual code implementation of the task is of little interest.

Like the control program, the task model maps sensory perception to robot motor response (see Figure 6.33). In order to obtain the model of a control program, the robot's sensory perception and its response to that perception is logged while it is executing the control program. Using the sensory perception as input and motor response as output, the same modelling technique as used in environment modelling (see Section 6.6) is used here in order to find a suitable model of the control program.

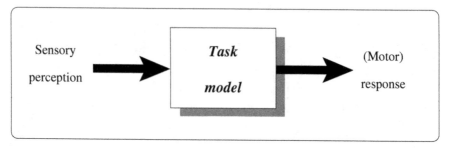

Figure 6.33. Task identification: a known function maps sensory perception to robot motor response

6.7.1 Task Identification: Identifying a Wall Following Behaviour Using ARMAX

This section presents a basic example of task identification in mobile robotics, determining a linear ARMAX model for the task of wall following. In order to make the fundamental mechanism of task identification clear, things have been kept simple: the amount of data used in this example is small, inputs and outputs are well correlated, and the model obtained is as uncomplicated as possible.

The objective of this example is to identify the relationship between sensory perception and motor response in a mobile robot that is performing a wall following task.

Irrespective of how this wall following behaviour was actually implemented on the robot, we will see whether a relationship can be determined between the robot's laser perceptions straight ahead ("laser 90") and 45° to the right ("laser 135") as inputs and the turning speed of the robot ($\dot{\phi}$) as output. This is depicted in Figure 6.34, the actual input and output values used in this example are given in Table 6.8 and shown in Figure 6.35.

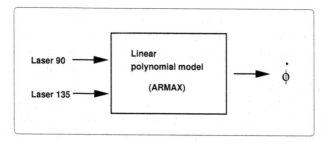

Figure 6.34. The identification task

One of the simplest Armax models conceivable is a model that uses no regression on the output at all (regression order zero on y), and a regression order

Table 6.8. Numerical values of the data shown in Figure 6.35 (read from left to right)

Laser 135 (input)

1.64 1.6 1.55 1.54 1.55 1.54 1.62 1.76 1.98 2.04
1.92 1.89 1.88 1.92 1.91 1.91 1.86 1.79 1.77 1.7
1.64 1.62 1.57 1.55 1.54 1.53 1.53 1.64 1.71 1.9
2.01 1.92 1.88 1.87 1.87 1.89 1.89 1.81 1.75 1.71
1.65 1.63 1.57 1.55 1.54 1.51 1.44 1.57 1.55 1.72
2.03 1.99 1.93 1.86 1.87 1.92 1.93 1.9 1.78 1.78
1.7 1.67 1.6 1.58 1.55 1.5 1.48 1.53 1.5 1.49
1.69 1.94 2.04 1.94 1.89 1.89 1.92 2.02 1.91 1.86
1.81 1.74 1.71 1.64 1.63 1.6 1.54 1.55 1.57 1.51
1.64 1.7 1.87 2.01 1.92 1.88 1.89 1.92 1.93 1.87
1.87 1.78 1.76 1.68 1.67 1.62 1.57 1.55 1.54 1.51
1.58 1.66 1.67 1.97 2.01 1.91 1.87 1.87 1.9 1.87
1.89 1.8 1.76 1.69 1.65 1.64 1.57 1.55 1.59 1.48
1.51 1.52 1.56 1.72 2.1 1.99 1.92 1.87 1.89 1.91
1.93 1.92 1.81 1.77 1.71 1.67 1.62 1.53 1.57 1.47
1.49 1.49 1.48 1.48 1.63 1.84 2.06 1.95 1.89 1.89
1.95 1.99 1.92 1.84 1.81 1.75 1.71 1.65 1.63 1.62
1.54 1.55 1.52 1.55 1.55 1.75 1.93 2.02 1.93 1.89
1.91 1.91 1.9 1.92 1.88 1.79 1.75 1.7 1.63 1.66
1.57 1.58 1.54 1.53 1.53 1.58 1.75 1.8 2.01 1.93

Laser 90 (input)

1.31 1.24 1.17 1.09 1.04 0.97 0.91 0.89 0.85 0.84
0.9 1.   1.12 1.45 1.66 1.76 1.7 1.61 1.52 1.44
1.37 1.3 1.22 1.16 1.09 1.02 0.96 0.92 0.88 0.85
0.87 0.91 0.98 1.19 1.37 1.67 1.77 1.7 1.6 1.52
1.44 1.37 1.29 1.21 1.16 1.08 1.   0.96 0.91 0.87
0.86 0.84 0.88 0.97 1.19 1.55 1.8 1.75 1.76 1.59
1.52 1.43 1.35 1.27 1.23 1.14 1.07 1.   0.95 0.89
0.85 0.83 0.82 0.85 0.92 1.14 1.37 1.84 1.77 1.68
1.61 1.53 1.44 1.38 1.3 1.23 1.16 1.09 1.03 0.96
0.92 0.87 0.84 0.86 0.89 1.   1.12 1.47 1.73 1.76
1.7 1.6 1.51 1.42 1.37 1.29 1.21 1.14 1.11 1.01
0.97 0.92 0.87 0.86 0.86 0.91 1.01 1.17 1.5 1.54
1.78 1.68 1.6 1.51 1.45 1.36 1.28 1.19 1.15 1.07
1.01 0.95 0.9 0.87 0.85 0.84 0.88 0.99 1.2 1.53
1.82 1.74 1.67 1.58 1.52 1.44 1.35 1.26 1.22 1.14
1.07 1.   0.95 0.89 0.85 0.82 0.81 0.85 0.9 1.07
1.47 1.86 1.77 1.7 1.62 1.54 1.45 1.39 1.31 1.25
1.17 1.1 1.03 0.97 0.93 0.88 0.86 0.84 0.89 1.
1.2 1.36 1.6 1.76 1.7 1.6 1.53 1.44 1.39 1.3
1.23 1.16 1.1 1.03 0.97 0.91 0.88 0.84 0.87 0.9

Rotational velocity (output)

0.05 0.06 0.07 0.07 0.08 0.1 0.13 0.15 0.18 0.21
0.23 0.22 0.2 0.11 0.07 0.01 0.03 0.05 0.03 0.05
0.06 0.06 0.07 0.07 0.08 0.09 0.11 0.12 0.15 0.18
0.21 0.22 0.22 0.17 0.13 0.05 0.   0.03 0.05 0.05
0.06 0.05 0.06 0.07 0.07 0.08 0.11 0.11 0.13 0.16
0.18 0.22 0.23 0.24 0.19 0.09 0.   0.   0.08 0.04
0.06 0.03 0.07 0.06 0.05 0.08 0.08 0.08 0.1 0.15
0.15 0.18 0.22 0.24 0.26 0.21 0.15 0.04 0.03 0.02
0.03 0.06 0.04 0.07 0.05 0.06 0.07 0.07 0.08 0.11
0.12 0.15 0.18 0.21 0.23 0.22 0.2 0.11 0.05 0.04
0.03 0.06 0.03 0.06 0.05 0.06 0.07 0.07 0.07 0.1
0.1 0.13 0.15 0.18 0.21 0.23 0.22 0.18 0.09 0.09
0.01 0.03 0.04 0.05 0.06 0.04 0.07 0.06 0.05 0.09
0.09 0.11 0.14 0.16 0.19 0.22 0.24 0.23 0.18 0.11
0.02 0.04 0.02 0.03 0.06 0.06 0.04 0.08 0.04 0.09
0.09 0.09 0.11 0.1 0.15 0.18 0.22 0.24 0.27 0.23
0.12 0.01 0.02 0.05 0.04 0.05 0.04 0.06 0.06 0.05
0.07 0.07 0.09 0.1 0.12 0.16 0.18 0.21 0.23 0.22
0.17 0.13 0.08 0.   0.02 0.05 0.05 0.05 0.07 0.04
0.07 0.06 0.08 0.09 0.11 0.13 0.15 0.19 0.2 0.22

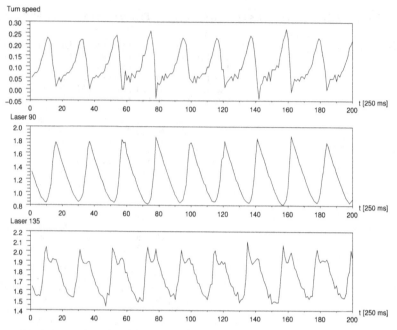

Figure 6.35. Input and output data used in the task identification example. The robot's motor response to sensory perception (the turning speed) is given in the top graph, the two laser perceptions used as input to the model (laser 90 and laser 135) are given below. The numerical values of all three graphs are shown in Table 6.8

of one on the input (*i.e.* only the inputs u at time t and $t - 1$ will be used to estimate the output y).

In Scilab this identification can be achieved as follows:

```
armax(0,1,[y';zeros(1:200)],u')

 A(x)  =

 !   1        0  !
 !               !
 !   0        1  !


 B(x)  =

 ! - 0.0060399 + 0.2257918x    - 0.1456451 - 0.0820220x  !
 !                                                        !
 !   0                           0                        !
```

This results in the model given in Equation 6.8:

$$\dot{\phi} = \tag{6.8}$$
$$-0.0060399l_{135}(t)$$
$$+0.2257918l_{135}(t-1)$$
$$-0.1456451l_{90}(t)$$
$$-0.0820220l_{90}(t-1)$$

with l_{90} and l_{135} the sensor readings obtained from the laser range finder of the robot in the direction of 90° (straight ahead) and 135° (45° to the right of the travelling direction), as shown in Figure 6.35. A brief note: instead of using the generic Scilab command `armax`, as we have done here, we could have used the program given in Section 6.4.3. Because that programme uses the first half of the data to obtain a model which is then validated against the second half of the data, the obtained model differs slightly from the one given in Equation 6.8, but is equally close to the original rotational velocity of the robot.

If we plot the actual rotation velocity $\dot{\phi}$ against the model given in Equation 6.8, we see that even this simple linear model actually provides a very good fit (Figure 6.36)!

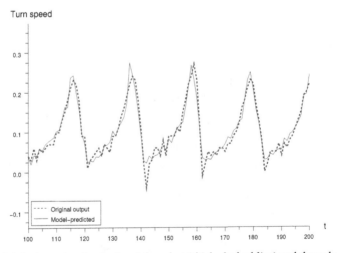

Figure 6.36. Original rotational velocity of the robot (*thick, dashed line*) and the polynomial model of it given in Equation 6.8 (*thin line*)

In fact, it is always worthwhile trying a linear Armax model for robot identification tasks; often they prove to be adequate for the modelling task at hand. For non-linear relationships, obviously, non-linear NARMAX models are needed, which are, however, more complicated to determine.

6.7.2 Task Identification: Identifying Wall Following Behaviour Using Narmax

Let us investigate the same behaviour — wall following — in a more complex environment. In this second example of task identification, a Magellan Pro mobile robot executed the wall following task in the environment shown in Figure 6.37.

Figure 6.37. The environment in which the wall following task was executed

The "original" wall following was achieved by using a back propagation neural network that had been trained to use sonar sensor input to produce the correct motor response, similar to the method discussed in [Iglesias et al., 1998]. The resulting "original" trajectory is shown in Figure 6.38.

We then identified the wall following task, using a NARMAX process, and obtained the model given in Table 6.9.

The inputs u1 to u16 of the model given in Table 6.9 were obtained by using the robot's sixteen sonar range readings, inverting them (so that short distances produce large values), and then setting all values of 0.25 to zero[2].

The next step, obviously, is to run the robot through the NARMAX model, rather than the original neural network. The resulting trajectory is shown in Figure 6.39.

Comparing Figures 6.38 and 6.39 clearly shows that both trajectories resemble each other well. The question of whether these two behaviours are *the same* is a difficult one, and discussed further below in Section 6.9.

---

[2] The Magellan's sonar sensors only return a valid range reading for distances up to 4m. Beyond that distance, "4m" is returned. By setting them to zero, we essentially remove all $\frac{1}{4} = 0.25$ readings.

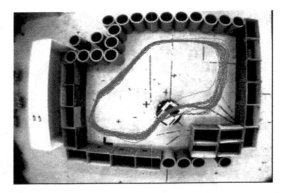

Figure 6.38. The wall following behaviour observed in the environment shown in Figure 6.37, using an artificial neural network controller

Table 6.9. The NARMAX model of the wall following behaviour. u specifies inputs 1 to 16 (see text for explanation), and n is the time step for which the rotational velocity r is being modelled

$$
\begin{aligned}
r(n) = \ & -0.3096661 \\
& -0.1243302 * u(n, 1) \\
& -0.0643841 * u(n{-}2, 1) \\
& -0.0389028 * u(n, 3) \\
& -0.1116723 * u(n, 9) \\
& +0.1749080 * u(n, 13) \\
& +0.0897680 * u(n, 14) \\
& -0.0541738 * u(n, 15) \\
& -0.0880687 * u(n, 16) \\
& +0.1128464 * u(n, 1)^2 \\
& +0.0789251 * u(n{-}2, 1)^2 \\
& +0.1859527 * u(n, 9)^2 \\
& -0.0202462 * u(n, 13)^2 \\
& +0.0531564 * u(n, 15)^2 \\
& +0.0996978 * u(n, 16)^2 \\
& +0.0608442 * u(n{-}1, 1) * u(n{-}1, 16) \\
& -0.0507206 * u(n{-}2, 1) * u(n{-}2, 9) \\
& +0.0283438 * u(n, 2) * u(n, 14) \\
& +0.0669943 * u(n, 2) * u(n, 16) \\
& -0.0519697 * u(n{-}1, 2) * u(n, 16) \\
& +0.0714956 * u(n, 3) * u(n{-}1, 16) \\
& +0.0534592 * u(n{-}1, 3) * u(n, 15) \\
& -0.0297800 * u(n, 13) * u(n, 14)
\end{aligned}
$$

6.7.3 Platform-Independent Programming Through Task Identification: The RobotMODIC process

The task identification scenario shown in Figure 6.33 determines the association between the robot's sensory perception and its motor response to that percep-

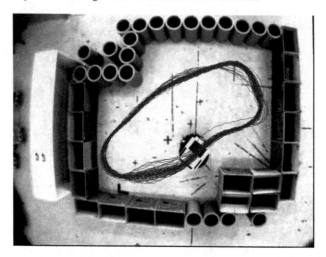

Figure 6.39. The trajectory observed when the robot was controlled by the NARMAX model

tion. In other words: it identifies the robot's control program. This relationship is expressed in a transparent and analysable function, such as for example the polynomial shown in Table 6.9. Whilst the original control program of the robot could be designed by *any* method available to robot engineers, be it control theory, machine learning techniques or any other methods, task identification leads to one unified expression of that very same task.

The obvious application that follows from this consideration is that the behaviour of one robot can be transferred to another robot, using task identification. This process, which we call RobotMODIC (Robot Modelling, Identification and Characterisation) is depicted in Figure 6.40.

The original coupling between perception and action (left hand side of Figure 6.40) is identified through ARMAX or NARMAX identification (right hand side of Figure 6.40). The resulting polynomial function is then used to determine motor responses to sensory perceptions (middle of Figure 6.40), resulting in comparable behaviour to that displayed by the robot, using the original controller.

The interesting point about this application of robot identification is that the "cloned" process can be run either on the original robot — providing an alternative and simpler way of programming to the original controller — or on another, even physically different robot (provided this second robot has access to similar sensor modalities as the original robot)! In effect, the RobotMODIC process provides a platform-independent method of robot programming.

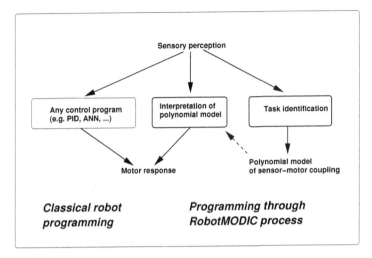

Figure 6.40. The RobotMODIC process

Experiment: "Robot Java"

That platform-independent robot programming through task identification works
in practice was demonstrated through the following experiment, which we conducted in collaboration with the Intelligent Systems Group at the University of
Santiago de Compostela: the "identified" behaviour of a wall following Magellan Pro robot was implemented on a Nomad 200 robot (resulting in a very short
control program for the Nomad, consisting essentially of one polynomial).

Although the robots differed, and the Magellan and the Nomad operated in
two different laboratories, the Nomad was able to execute wall following behaviour successfully in a real world environment in Santiago, without ever fail-
ing for more than one hour of operation. While this is only an existence proof, it
nevertheless demonstrates that cross-platform robot programming through robot
identification is possible.

6.7.4 "Programming" Through Training: Door Traversal using the RobotMODIC Process

The "translation" of robot control software from one robot platform to another,
using the RobotMODIC process, is a convenient and very fast way of achieving
similar robot behaviour across many different types of robots. It still has one
weakness, though: in order to identify the robot's behaviour and express it in the
form of a linear or nonlinear polynomial, the behaviour must first be present in
the robot. In practice, this means that we have to write robot control code by
traditional means first in order to re-express it in the form of polynomials.

There is a way round this expensive process: robot training. If we guide the robot manually through the required sensor-motor task and identify *that* behaviour, we bypass the traditional programming process and arrive at the polynomial representation of the task more or less immediately.

The following case study shows how this is done in the case of door traversal, a fairly complex sensor-motor task that requires fine control of the robot's motion, and the use of different sensors at different stages of the motion (the laser sensor only faces forward, and cannot be used once the robot is actually traversing the opening. At that point, sonar or infrared sensors have to be used).

Figure 6.41 shows the experimental setup. During both the training phase and the autonomous motion phase after task identification the robot started somewhere within the shaded area to the left of the door. The door itself had a width of two robot diameters, which required good motion control on the side of the robot in order not to get stuck. The robot used was again the Magellan Pro *Radix* shown in Figure 1.1.

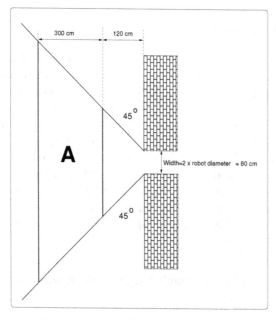

Figure 6.41. Experimental scenario for the door traversal behaviour. Starting positions of the robot were within the trapezoid area labelled "A"

To obtain training data for the RobotMODIC process, the robot was then taken to an arbitrary starting position within the area "A" shown in Figure 6.41, and manually driven through the door. All sensor and motor signals were logged at a sampling rate of 4Hz during this operation. This process of gathering data was repeated 39 times, resulting in one time series of 39 joined sections of data, each section consisting of several hundred data points.

This training data was then used to identify the task of door traversal, resulting in the nonlinear polynomial shown in Table 6.10, where d indicates averaged laser range readings over sectors of $15°$, and s denotes sonar range readings.

Table 6.10. NARMAX model of the angular velocity $\dot{\theta}$ for the door traversal behaviour, as a function of laser and sonar information d and s

$$\dot{\theta}(t) =$$
$$+0.272$$
$$+0.189 * (1/d_1(t))$$
$$-0.587 * (1/d_3(t))$$
$$-0.088 * (1/d_4(t))$$
$$-0.463 * (1/d_6(t))$$
$$+0.196 * (1/d_8(t))$$
$$+0.113 * (1/d_9(t))$$
$$-1.070 * (1/s_9(t))$$
$$-0.115 * (1/s_{12}(t))$$
$$+0.203 * (1/d_3(t))^2$$
$$-0.260 * (1/d_8(t))^2$$
$$+0.183 * (1/s_9(t))^2$$
$$+0.134 * (1/(d_1(t) * d_3(t)))$$
$$-0.163 * (1/(d_1(t) * d_4(t)))$$
$$-0.637 * (1/(d_1(t) * d_5(t)))$$
$$-0.340 * (1/(d_1(t) * d_6(t)))$$
$$-0.0815 * (1/(d_1(t) * d_8(t)))$$
$$-0.104 * (1/(d_1(t) * s_8(t)))$$
$$+0.075 * (1/(d_2(t) * s_7(t)))$$
$$+0.468 * (1/(d_3(t) * d_5(t)))$$
$$+0.046 * (1/(d_3(t) * s_5(t)))$$
$$+0.261 * (1/(d_3(t) * s_{12}))$$
$$+1.584 * (1/(d_4(t) * d_6(t)))$$
$$+0.076 * (1/(d_4(t) * s_4(t)))$$
$$+0.341 * (1/(d_4(t) * s_{12}(t)))$$
$$-0.837 * (1/(d_5(t) * d_6(t)))$$
$$+0.360 * (1/(d_5(t) * d_7(t)))$$
$$-0.787 * (1/(d_6(t) * d_9(t)))$$
$$+3.145 * (1/(d_6(t) * s_9(t)))$$
$$-0.084 * (1/(d_6(t) * s_{13}(t)))$$
$$-0.012 * (1/(d_7(t) * s_{15}(t)))$$
$$+0.108 * (1/(d_8(t) * s_3(t)))$$
$$-0.048 * (1/(d_8(t) * s_6(t)))$$
$$-0.075 * (1/(d_9(t) * d_{16}(t)))$$
$$-0.105 * (1/(d_{10}(t) * d_{12}(t)))$$
$$-0.051 * (1/(d_{10}(t) * s_{12}(t)))$$
$$+0.074 * (1/(d_{11}(t) * s_1(t)))$$
$$-0.056 * (1/(d_{12}(t) * s_7(t)))$$

Figure 6.42 shows the trajectories generated by the human operator (top), which were used to generate the training data for the RobotMODIC process, and the trajectories generated when the robot moved autonomously, controlled by the polynomial given in Table 6.10.

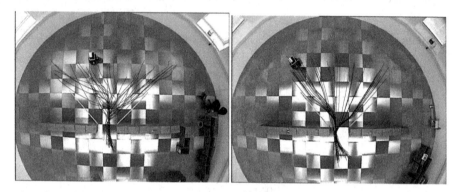

Figure 6.42. *Left*: The robot's trajectories under manual control (39 runs, training data). *Right*: Trajectories taken under model control (41 runs, test data)

In both cases the robot moved through the opening without problems. Careful analysis of the trajectories obtained under autonomous control reveals that the robot moved a lot more smoothly and centrally through the door than when driven by the human!

In summary, this case study of door traversal demonstrated that it is possible to obtain precise and relatively complex motion control through robot training and subsequent task identification, a process that is far more efficient, easier and faster than programming a robot in the traditional way.

6.8 Sensor Identification

The platform-independent programming of robots discussed in Section 6.7.3 requires, of course, that both the originally used robot ("robot A") and the robot that uses the identified function (the polynomial) as a controller ("robot B") use the same type of sensors. There is a problem, however, if robot A used, say, sonar sensors to produce the original behaviour, and robot B only possessed laser sensors. The function identified to map sonar perception to motor response cannot be used on robot B in this case.

Or can it?

Many robot sensor modalities, in particular range sensors, generate comparable signals. For example, while not identical, laser range finders and sonar range finders can both generate similar depth maps of objects in front of the robot. If we had a "translation" from laser to sonar, we *could* use the RobotMODIC process to translate the original behaviour, executed on robot A, to an equivalent behaviour on robot B, even though both robots use different sensor modalities. We refer to this method as "sensor identification', the process is shown in Figure 6.43.

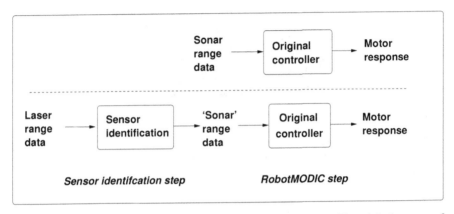

Figure 6.43. Sensor identification used in the RobotMODIC process. The original process of behaviour generation is shown at the *top*. The *bottom* depicts the behaviour generation process, using sensor identification through the RobotMODIC process

Sensor Identification: Example

Such sensor identification is indeed possible in certain cases. In the following example, we have logged sonar and laser perceptions of a Magellan Pro mobile robot, as it was moving randomly in a real world environment. The robot's trajectory is shown in Figure 6.44.

The sensor identification task we were interested in here is that of translating laser perception into sonar perception, as indicated in Figure 6.43. To achieve this, we used the range signals of five laser sensors as inputs, and generated one sonar perception as output. This is shown in Figure 6.45.

The polynomial NARMAX model obtained is shown in Table 6.11, and the correspondence between actual sonar perception and the "sonar" perception predicted, using five laser range measurements, is shown in Figure 6.46.

6.9 When Are Two Behaviours the Same?

6.9.1 Static Comparison Between Behaviours

Earlier in this chapter (Section 6.7.2) we identified the task of wall following, originally implemented through an artificial neural network, using a NARMAX

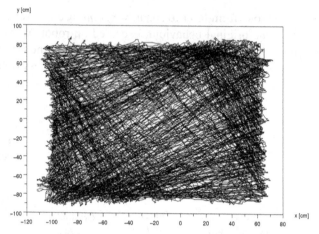

Figure 6.44. Trajectory of a randomly-moving Magellan Pro. While the robot was following this trajectory, both the robot's sonar and laser perceptions were logged for subsequent sensor identification

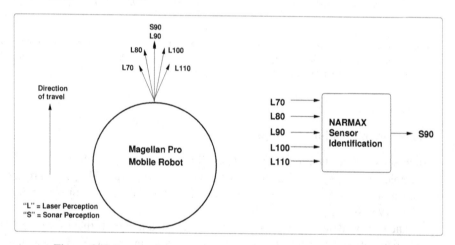

Figure 6.45. Input and output data used in the sensor identification example

model, and then executed the model on the robot. The trajectory achieved with the original controller is shown in Figure 6.38, the one achieved with the NAR-MAX model in Figure 6.39. Both trajectories are shown together in Figure 6.47[3]. They do look similar, but following the principles of scientific robotics we would like to *quantify* that similarity.

---

[3] Note that the coordinate values of both trajectories have been normalised in such a way that the geometrical relationships have been preserved, by dividing *all* coordinate values by $max(x_{ANN}, y_{ANN}, x_{Model}, y_{Model})$.

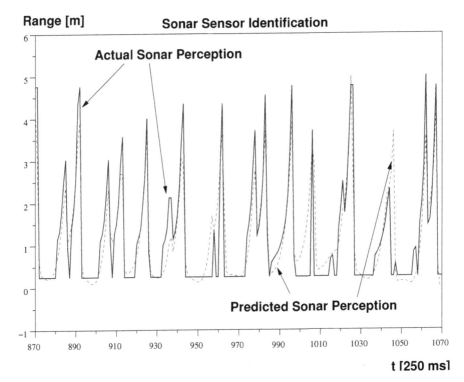

Figure 6.46. Actual sonar perception logged (*thick solid line*) *vs* sonar perception predicted by the model given in Table 6.11 (*dashed thin line*)

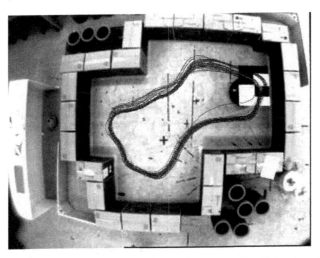

Figure 6.47. The trajectory of the ANN wall following controller (*faint dotted line*) *vs* the trajectory of the NARMAX polynomial controller (*solid line*)

Table 6.11. Translation of laser perceptions L70 to L110 into sonar perception S90. See Figure 6.45 for an explanation of the symbols used

$$
\begin{aligned}
S90(t)= & -0.09486578751437843571 \\
& +1.34090710655649880678 * L70(t) \\
& -1.46993209143653791315 * L70(t\text{-}1) \\
& -0.10133156027639933505 * L70(t\text{-}2) \\
& +1.62510983787471263717 * L80(t) \\
& -0.04077783307865779500 * L80(t\text{-}2) \\
& -2.58382888545996491914 * L90(t) \\
& +1.34797456078995359086 * L100(t) \\
& -0.30776670071816458751 * L110(t\text{-}1) \\
& -1.24584261624820435976 * L70(t)^2 \\
& +0.12259678893997662252 * L70(t\text{-}1)^2 \\
& -3.70451454646306554963 * L80(t)^2 \\
& -0.05547179821486561413 * L80(t\text{-}1)^2 \\
& -1.54497725705582955591 * L90(t)^2 \\
& +0.08003594836197346074 * L100(t)^2 \\
& -0.11219782488872127868 * L110(t)^2 \\
& -0.22856361726059648554 * L110(t\text{-}1)^2 \\
& +1.13094692355659165450 * S90(t\text{-}1) \\
& -0.10649893253405356974 * S90(t\text{-}1)^2 \\
& +1.70065894822170871059 * L70(t) * L80(t) \\
& -0.17397893514679030336 * L70(t) * L80(t\text{-}2) \\
& +0.16703072280923750292 * L70(t) * L90(t) \\
& +0.24576961719240705828 * L70(t) * S90(t\text{-}1) \\
& +0.51897979560886331463 * L70(t\text{-}1) * L80(t\text{-}2) \\
& +0.67296902266249047919 * L70(t\text{-}1) * L90(t) \\
& -0.21876283482185332474 * L70(t\text{-}1) * u(n\text{-}2, 4) \\
& -0.08831850086310211179 * L70(t\text{-}1) * S90(t\text{-}1) \\
& +0.21165059817172712786 * L70(t\text{-}2) * L110(t) \\
& -0.20194679975663892835 * L70(t\text{-}2) * S90(t\text{-}1) \\
& +4.51797017772498765709 * L80(t) * L90(t) \\
& -0.42664008070691378238 * L80(t) * S90(t\text{-}1) \\
& -0.36439534395116823795 * L80(t\text{-}1) * L100(t) \\
& -0.21557583313895936628 * L80(t\text{-}2) * L100(t\text{-}1) \\
& +0.54408085200311495644 * S90(t\text{-}1) * L110(t\text{-}1) \\
& +0.30411288763928323586 * L90(t\text{-}2) * L100(t\text{-}1) \\
& -0.11689726904905589633 * L110(t\text{-}1) * S90(t\text{-}1)
\end{aligned}
$$

One way to say something about the similarity or dissimilarity of the two trajectories shown in Figure 6.47 would be to treat both trajectories essentially as images. If there is no significant difference between the distribution of the "pixels" (x and y positions of the robot) in original and modelled behaviour, we could argue that there is at least strong evidence that both behaviours have fundamental properties in common. Clearly, this is a static test, and the behaviour of a robot is very much governed also by dynamics, but nevertheless this is a first attempt at measuring statistically whether two trajectories are similar or not.

The distributions of the robot's x and y coordinates for the original behaviour and the model are shown in Figure 6.48.

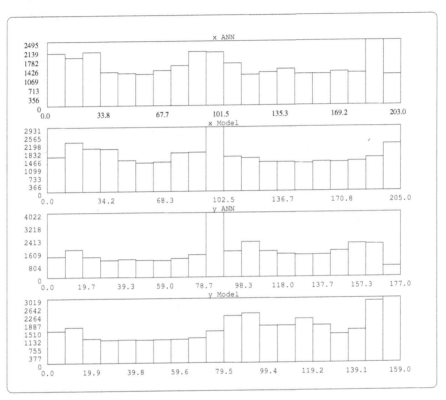

Figure 6.48. The occupancy probabilities for x and y position of the robot, both for the original ANN wall follower and the NARMAX model

Using the Mann-Whitney U-test discussed in Section 3.4.2, we find that there is no significant difference ($p<0.05$) between the two distributions of the x position of the robot. For the y coordinate distribution, however, the difference between model and original *is* significant. This is also visible qualitatively from Figure 6.47. This implies that the occupancy of space generated by the modelled behaviour is not significantly different from the original along the x axis, but that along the y axis the model-driven robot occupies space differently to the ANN-driven robot. Incidentally, if we scale all trajectories so that they occupy the interval [0 1], *i.e.* introduce a distortion and only consider the geometrical shape of trajectories, the significant difference between the two y distributions disappears.

This comparison of two trajectories is static, because it only takes into account the physical positions the robot has occupied while executing its control program. The interaction between robot and environment, however, has an important dynamic aspect as well. It does matter, at least in certain applications, whether the robot moved fast or slowly, whether it turned fast or slowly, whether it stopped often, *etc.* These dynamics are not captured by analysing physical occupancy, but they can be analysed dynamical systems theory.

6.9.2 Comparison of Behaviour Dynamics

In Chapter 4 we discussed that the behaviour of a dynamical system, such as an autonomous mobile robot, can be described either in physical space, or in the system's phase space. The latter has the advantage that a number of tools and quantitative measures are available, and we'll apply the techniques presented in Section 4.4.2 (prediction horizon) and Section 4.4.5 to the two trajectories shown in Figure 6.47.

Prediction Horizon

Figure 6.49 shows the prediction horizon for the robot's movement along the x-axis, using the original ANN controller.

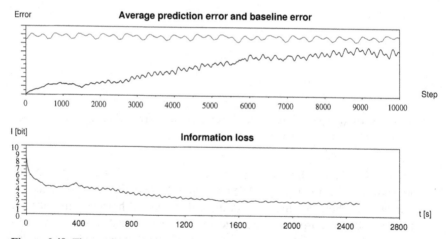

Figure 6.49. The prediction horizon computed for the robot's movement along the x-axis in the original (ANN) wall follower

Figure 6.49 shows clearly that the robot's wall following behaviour is extremely predictable. Even if we try to predict the robot's x position 10000 time steps ahead (that is over 40 min), we get a far smaller prediction error using the robot's past position than using a random point from the first half of our data as our prediction. This means that this behaviour essentially is deterministic, and very predictable. The Lyapunov exponent is essentially zero.

Figure 6.50 shows the same computation for the prediction of the robot's x position when it executes the NARMAX model of wall following.

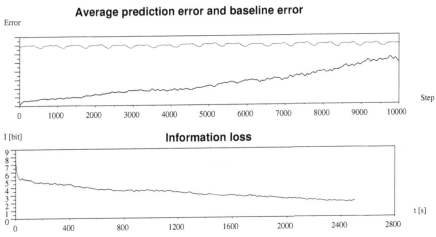

Figure 6.50. The prediction horizon computed for the robot's movement along the x-axis in the identified NARMAX model wall follower

The results are similar to those obtained using the original behaviour: Even for predictions a long time ahead the data-based prediction outperforms the prediction using the baseline. Again, the Lyapunov exponent is essentially zero, and indistinguishable from the Lyapunov exponent for the ANN wall follower.

Using Entropy

There is another way of quantifying similarities or differences between the original behaviour and the one achieved with the model, incorporating both static and dynamical properties: looking at the correlation between original and modelled behaviour.

Figure 6.51, which shows 1000 s of $x_{ANN}(t)$ and $x_{Model}(t)$, reveals that the robot moves along the x-axis in a similar manner in both cases, but faster in the case of $x_{Model}(t)$. The two time series are sometimes in phase, sometimes out of phase.

This fact can be seen also from the cross correlation of $x_{ANN}(t)$ and $x_{Model}(t)$, shown in Figure 6.52.

A contingency table analysis will reveal the degree of correlation between $x_{ANN}(t)$ and $x_{Model}(t)$. For 10000 data points, the table is shown in Table 6.12.

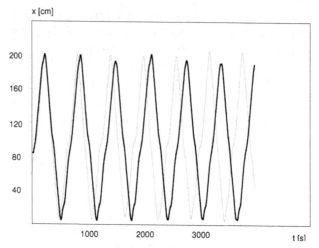

Figure 6.51. 1000 s of robot movement along the x axis, for original (*thick lines*) and modelled behaviour (*thin lines*). It is clearly visible that the model has a higher frequency

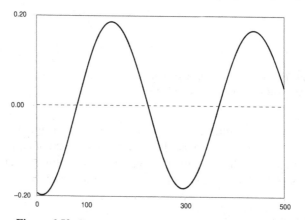

Figure 6.52. Cross correlation of $x_{ANN}(t)$ and $x_{Model}(t)$

Table 6.12. Contingency table, relating $x_{orig}(t)$ to $x_{Model}(t)$

57. 84. 61. 47. 172. 83. 58. 42. 24. 1.
51. 29. 26. 32. 104. 8. 14. 18. 23. 29.
58. 35. 12. 41. 61. 24. 4. 8. 16. 48.
57. 26. 61. 37. 37. 26. 33. 17. 31. 83.
88. 88. 32. 29. 60. 33. 33. 72. 78. 43.
57. 45. 13. 16. 36. 39. 48. 29. 21. 37.
74. 0. 22. 19. 33. 59. 49. 20. 20. 41.
49. 24. 1. 48. 48. 29. 25. 37. 33. 30.
29. 23. 36. 57. 40. 18. 30. 38. 54. 27.
 5. 28. 60. 55. 29. 43. 37. 37. 43. 75.

As can already be seen from Figure 6.51, however, the two trajectories have very similar shape, but different periods. For the first 1000 data points, the two trajectories are very much in phase, but by data point 3500 they are in anti phase! Any contingency table analysis in this case, therefore, will be dependent upon the number of data points considered, and indeed, the uncertainty coefficient will decrease from 0.30 for 1000 data points to 0.01 for 14000 data points. A χ^2 analysis (Section 3.8.1) shows that there is a significant correlation between $x_{orig}(t)$ and $x_{Model}(t)$, be it for 1000 or 14000 data points, but the correlation becomes very small over long period of time. This indicates that there are dynamical differences between the original model and the model, although both produce very similar trajectories and have similar Lyapunov exponents.

6.10 Conclusion

6.10.1 Summary

A common modelling scenario in mobile robotics is to model input-output relationships, for example:

- Robot position *vs* sensory perception ("simulation")
- Sensory perception *vs* robot position ("self-localisation")
- Sensory perception *vs* motor response ("robot programming")
- Sensory perception *vs* sensory perception ("sensor signal translation")

This chapter demonstrated that such models can be obtained, for example, by using artificial neural networks (Section 6.3). The disadvantage of this approach, particularly in view of the objectives of scientific mobile robotics to analyse robot behaviour, is that ANN models are opaque, so that the relationship between input and output, although modelled, nevertheless remains largely unknown.

We therefore presented an alternative, which yields transparent models that can be analysed mathematically: linear or nonlinear models, using polynomials (ARMAX and nonlinear ARMAX modelling).

Identifying input-output relationships is commonly referred to as system identification; correspondingly we refer to its application to robots as robot identification. Examples given in this chapter show that through robot identification it is possible to build faithful simulators of robot-environment interaction (Section 6.6), that robot self localisation is possible (Section 6.6.3), that cross-platform programming ("Robot Java") is possible (Section 6.7), and that signals from one sensor modality can be "translated" into another sensor modality (Section 6.8). These techniques provide a powerful set of tools for the development, analysis and refinement of robot control code.

6.10.2 Open Questions

In all modelling tasks, the big question is whether the obtained model is faithful to the original, *i.e.* whether it retains the fundamental properties of the original, despite the abstractions and generalisations employed.

This is a difficult question to answer, and largely dependent upon the definition of "faithful". In Section 6.9 some examples were given of how two robot behaviours can be compared, by i) comparing the static appearance of the robot's trajectories, and ii) comparing the dynamics of the robot's motions in the two cases. These are certainly not the only ways to compare behaviours, and other ways of comparing behaviour are conceivable.

If two output streams — the original and its model — are to be compared, the correlation between original and model can be determined and analysed for significance. In this chapter this was done for some sensor modelling tasks, using correlation coefficients such as the Spearman rank correlation coefficient r_S. Also, contingency table analysis (χ^2, Cramer's V and entropy-based measures) can be used to determine the significance of a correlation between original and model.

7

Conclusion

Summary. The final chapter summarises the material presented in this book, and draws some conclusions. It also points to open questions and outstanding issues in mobile robotics research.

7.1 Motivation

Mobile robotics is continuously gaining in importance in science and industry. On the one hand, this is due to a widening range of industrial applications, ranging from the by now commonplace transportation, inspection, surveillance and cleaning tasks, to niche applications such as hazardous materials handling, or work in unstructured environments such as disaster sites. On the other hand, mobile robots are gaining influence in the behavioural sciences as tools to investigate the foundations of behaviour, and to gain a better insight into the relationship between robot, task and environment — in other words, which parameters of these three categories produce which behaviour? It is the latter that this book is mainly concerned with. The first reason for the work discussed in this book, therefore, is this question:

- What is the relationship between robot hardware, robot software and the environment on the one hand, and the robot behaviour resulting thereof on the other hand? This point is sometimes referred to as "theory of robot-environment interaction"

There is a second motivation behind this book. Because of a current lack of scientific tools and ways to describe behaviour, much of mobile robotics research to date is confined to the presentation of existence proofs. Unlike other natural sciences, mobile robotics research does not commonly use independent replication and confirmation of results, mostly because we haven't got the "language" to present our results in such a way that such replication and verification can easily be achieved.

A research practice that goes beyond existence proofs — the once-off presentation of results that demonstrates that something can be done, but not how something can be done *in general* — and that uses independent replication and verification of experiments, requires *quantitative*, measurable descriptions of results. The second motivation behind this book, therefore, is expressed in this question:

- How can robot-environment interaction be described quantitatively, *i.e.* measured, to allow replication and verification of results?

Satisfactory answers to these two questions would constitute a step towards a new, enhanced research practice in robotics. We would gain understanding not only of the "how?", but also of the "how in general?", *i.e.* we would be able to identify fundamental principles that govern robot-environment interaction, and exploit these to design task-achieving robots based on theoretical understanding, rather than trial and error. We would be able to make predictions about robot behaviour, and to analyse safety and stability of robot operation, based on theoretical understanding.

Mobile robotics will always be a discipline predominantly based on experiments with real robots. Even the best model will not replace the real thing, but through theoretical understanding of robot-environment interaction and quantitative descriptions this experimentation will become more focused and efficient.

7.2 Quantitative Descriptions of Robot-Environment Interaction

There are two major aspects to the interaction of a mobile robot with its environment, both of which we would like to measure: static aspects, such as the comparison between two behaviours in space, rather than over time, and dynamic aspects, which reflect that the robot's behaviour is a function of space *and* time.

7.2.1 Static Quantitative Descriptions of Robot-Environment Interaction

One typical scenario in mobile robotics research is the comparison between two or more solutions to the same problem. For example, in robot self-localisation one might be interested whether a landmark-based localisation performs better than one that is based on dead reckoning. What does "better" mean in this case? Provided the performance of both systems can be logged in some way — for the localisation scenario constructing contingency tables that relate true position to perceived position (see Section 3.8.1) is a suitable method — statistical methods give us the means to compare performances quantitatively and to establish whether or not there is a statistically significant difference between them.

Section 3 presented a wide range of methods, commonly used in biology and psychology, that are applicable to mobile robotics. These methods establish whether two distributions differ or not (Section 3.3 for normally distributed data, Section 3.4 for data of any distribution), whether a sequence is random or not (Section 3.5), if there is a correlation between two sets of numerical data (sections 3.6 and 3.7), or if there is a correlation between two sets of categorical data (Section 3.8).

All of these tests are tests of comparison, where the behaviour of the robot in one situation (one particular manifestation of the triple robot-task-environment) is compared with the robot's behaviour if one element of robot, task or environment has been modified (for example by changing the robot's control code, *i.e.* the task). These tests do not take the dynamics of robot-environment interaction into account.

7.2.2 Quantitative Descriptions of the Dynamics of Robot-Environment Interaction

There is a second kind of quantitative description we wish to obtain to describe the robot's behaviour, capturing the *dynamics* of its interaction with the environment. One way to achieve this is to use dynamical systems theory, as discussed in Chapter 4.

A mobile robot, interacting with its environment, is a dynamical system, *i.e.* a system whose behaviour has to be described by taking time into account. In other words, the interaction of a mobile robot with the environment can be described by differential equations.

The tools discussed in Section 4 were originally developed to describe physical systems whose behaviour is governed by differential or difference equations, within a discipline now known as deterministic chaos. This term refers to the fact that these systems are not stochastic, but nevertheless are only partially predictable and behave, in some circumstances, as if they were indeed random.

Deterministic chaos methods are only applicable to signals that are deterministic, and stationary. Having established those two facts first, the analysis of the dynamics of robot-environment interaction then typically begins by reconstructing the robot's phase space through time lag embedding (Section 4.2.1).

Once the phase space has been reconstructed, it can then be described quantitatively, for instance by estimating the Lyapunov exponent (Section 4.4), the attractor's correlation dimension (Section 4.5) or the prediction horizon beyond which the system becomes unpredictable (Section 4.4.2).

7.3 A Theory of Robot-Environment Interaction

The ultimate goal of mobile robotics research, as discussed in this book, is to develop a coherent body of hypothetical, conceptual and pragmatic generalisations and principles that form the general frame of reference within which mobile robotics research is conducted — a "theory of robot-environment interaction".

Such a theory would provide two very useful benefits:

1. A theory will allow the formulation of hypotheses for testing. This is an essential component in the conduct of "normal science" [Kuhn, 1964].
2. A theory will make predictions (for instance regarding the outcome of experiments), and thus serve as a safeguard against unfounded or weakly supported assumptions.

In other words, a theory contains, in abstraction and generalisation, the essence of what it is that the triple of robot-task-environment does. This generalisation is essential: it highlights the important aspects of robot-environment interaction, while suppressing unimportant ones. Finally, the validity of a theory (or otherwise) can then be established by evaluating the predictions made applying the theory.

What form could such a theory take? Clearly, it could be expressed in the form of mathematical descriptions (formulae) of the relationships between variables that describe the robot's behaviour, like the relationship between force, mass and acceleration (Newton's law). If the value of some variables is known, then the remaining variables can be predicted.

Because a theory should generate testable and falsifiable hypotheses [Popper, 1959], there is another way of establishing a theory: rather than trying to unravel the intricate interaction between robot and environment, and attempting to express all aspects of it in mathematical formulae, one could construct computer models of the robot's interaction with the environment, and use these to generate testable hypotheses. This method is discussed in Chapter 6.

If the computer model is to capture the essence of the robot's behaviour, it must be generated from real data, rather than from theoretical considerations, which would be based on a theory that doesn't exist yet. In Chapter 6 we present such a method. Based on the established methods of system identification [Eykhoff, 1974, Ljung, 1987], we construct mathematical relationships between variables that govern the robot's behaviour, using a process we refer to as "robot identification". The benefits of this process are manifold:

- Robot identification expresses relationships in transparent functions that can be analysed by established methods
- Robot identification allows the transfer of control code between robot platforms, without the need of rewriting code for a different robot
- It allows the simple modification of control code, replacing one sensor modality with another without having to rewrite the entire control code
- It allows the construction of faithful robot simulators that support off-line design of control code
- It allows the *accurate* comparison of two different control mechanisms on the simulator, because the underlying model of robot-environment interaction remains unchanged
- Robot identification makes testable predictions about the robot's behaviour.

7.4 Outlook: Towards Analytical Robotics

Mobile robotics is a very practical discipline, a discipline concerned with building machines that will carry out real tasks in the real world. Whether these tasks are "factory style" tasks such as transportation and cleaning or "science style" tasks such as learning and autonomous navigation is immaterial for the discussion here; in all cases a coupling between perception and action has to be established through the designer-supplied robot control code. As in all engineering tasks, design benefits from analytical understanding, and it is one of the aims of this book to add this aspect to the current robotics research agenda.

In contrast to experimental mobile robotics — the dominant aspect of robotics research to date — scientific mobile robotics has the following characteristics (Section 2.8):

- Experimental design and procedure are guided by testable, falsifiable hypotheses, rather than based on the researcher's personal experience (i.e. on a "hunch")
- Experimental design and procedure are "question-driven", rather than "application-driven"
- Results are measured and reported quantitatively, rather than qualitatively
- Experimental results are replicated and verified independently (for example by other research groups), rather than presented as stand-alone existence proofs

This approach to experimentation with dynamical systems — not confined to mobile robots, but also relevant to the behaviour of animals or technical systems — rests on three pillars:

1. Description,
2. Modelling, and
3. Analysis.

Description in a scientific context must mean *quantitative* description of experimental results, *i.e. measurable* behaviour indicators that allow precise comparison between different experimental observations. Statistical methods like those discussed in Chapter 3 and descriptors of behaviour dynamics like those discussed in Chapter 4 can serve this purpose.

Modelling is the second pillar, because a model that captures the essence of the modelled behaviour in abstraction provides a focus, a condensed representation of those aspects that matter, omitting those that do not matter. It simplifies the understanding of the system under investigation. Methods like the ones discussed in Chapter 6 are one step towards achieving this goal.

Once a model of the system under investigation is obtained, we would like to know

- Is the model accurate, *i.e.* faithful to the modelled system?
- What does the model mean, *i.e.* how are we to interpret the model, and what do we learn from the model about the behaviour of the original system?
- Does the model lead to new insights?

Analysis, therefore, comprises the comparison between the behaviour of the original system and its model. In Section 6.9 some static and dynamic techniques were presented that address this issue, but here more work is needed to clarify what precisely we mean by "identical" behaviours.

Some interpretation of a model can be achieved, for instance, through sensitivity analysis. Recent work conducted at Essex shows that methods like the one discussed by [Sobol, 1993] can be used to quantify the importance of individual model terms, differentiating between major and minor model components. And such model analysis can indeed lead to new insights. For example, only by modelling the door traversal behaviour of our mobile robot, and subsequently analysing the model through sensitivity analysis did we realise that in fact the robot only used the sensors on its right hand side to go through a door — a completely unexpected result for a "symmetrical" task such as driving through a doorway. The benefits of theoretical understanding — in this case represented by a computer model of the robot's operation — are illustrated by this example: having established that a task (*e.g.* door traversal) requires only certain sensors, the robot's hardware and software can be simplified, resulting in cheaper robots that have to perform less computation.

Description, modelling and analysis, then, are the three pillars of research presented in this book, which attempts to define a new research agenda in mobile robotics. We are only at the beginning, though, and the examples and case studies given in this book are just one possible approach to tackle the issues. Future work will have to deepen the treatment of problems like the identity of behaviours, the "meaning" of models, the relationship between the operation of physical agents and their simulations and the theoretical limitations of computer modelling (see [Oreskes et al., 1994] for a discussion of this specific issue).

This book, therefore, is an invitation for discussion and further development, to refine and focus our research and experiments further in this emerging and exciting new area of Analytical Robotics.

References

[Abarbanel, 1996] Abarbanel, H. (1996). *Analysis of observed chaotic data*. Springer Verlag, New York.

[ANS, 2003] ANS (2003). *Tools for dynamics.* Applied Nonlinear Sciences, http://www.zweb.com/apnonlin.

[Arkin, 1998] Arkin, R. (1998). *Behavior-based robotics*. MIT Press, Cambridge, Mass.

[Arkin and Hobbs, 1992] Arkin, R. C. and Hobbs, J. (1992). Dimensions of communication and social organization in multi-agent robotics systems. In *From animals to animats 2*, Cambridge MA. MIT Press.

[Bacon, 1878] Bacon, F. (1878). *Novum Organum*. Clarendon Press, Oxford.

[Baker and Gollub, 1996] Baker, G. and Gollub, J. (1996). *Chaotic Dynamics*. Cambridge University Press, Cambridge, UK.

[Barnard et al., 1993] Barnard, C., Gilbert, F., and McGregor, P. (1993). *Asking questions in biology*. Longman Scientific and Technical, Harlow, UK.

[Bendat and Piersol, 2000] Bendat, J. S. and Piersol, A. G. (2000). *Random data : analyis and measurement procedures*. Wiley, New York.

[Beni and Wang, 1989] Beni, G. and Wang, J. (1989). Swarm intelligence in cellular robotic systems. In *Nato advanced workshop on robotics and biological systems*, Il Ciocco, Italy.

[Box et al., 1994] Box, G., Jenkins, G., and Reinsel, G. (1994). *Time Series Analysis*. Prentice-Hall.

[Braitenberg, 1987] Braitenberg, V. (1987). *Vehicles*. MIT Press, Cambridge, Mass.

[Burgard et al., 1998] Burgard, W., Cremers, A. B., Fox, D., Hähnel, D., Lakemeyer, G., Schulz, D., Steiner, W., and Thrun, S. (1998). Experiences with an interactive museum tour-guide robot. *Artificial Intelligence*, 114:3–55.

[Chen and Billings, 1989] Chen, S. and Billings, S. A. (1989). Representations of non-linear systems: The narmax model. *Int. J. Control*, 49:1013–1032.

[Critchlow, 1985] Critchlow, A. (1985). *Introduction to Robotics*. Macmillan, New York.

[Demiris and Birk, 2000] Demiris, J. and Birk, A., (Eds.) (2000). *Interdisciplinary Approaches to Robot Learning*. World Scientific Publishing.

[Dorigo and Colombetti, 1997] Dorigo, M. and Colombetti, M. (1997). *Robot Shaping: An Experiment in Behavior Engineering*. MIT Press.

[Dudek and Jenkin, 2000] Dudek, G. and Jenkin, M. (2000). *Computational principles of mobile robotics*. Cambridge University Press, Cambridge, UK.

[Dunbar, 2003] Dunbar, K. (2003). Scientific thought. In Nadel, L., (Ed.), *Encyclopedia of Cognitive Science*, London. Nature Publishing Group.

[EXN, 2003] EXN (2003). *Discovery channel.* http://www.exn.ca/Stories/2003/-03/11/57.asp.

[Eykhoff, 1974] Eykhoff, P. (1974). *System identification : parameter and state estimation.* Wiley, New York.

[Franklin, 1996] Franklin, J., (Ed.) (1996). *Recent Advances in Robot Learning.* Kluwer Academic Publishers.

[Fraser and Swinney, 1986] Fraser, A. M. and Swinney, H. L. (1986). Independent coordinates for strange attractors from mutual information. *Physical Review A*, 33:1134–1140.

[Fuller, 1999] Fuller, J. L. (1999). *Robotics : introduction, programming, and projects.* Prentice-Hall.

[Gillies, 1996] Gillies, D. (1996). *Artificial intelligence and scientific method.* Oxford University Press.

[Gower, 1997] Gower, B. (1997). *Scientific method : a historical and philosophical introduction.* Routledge, London.

[Harris, 1970] Harris, E. (1970). *Hypothesis and perception.* George Allen and Unwin Ltd.

[Iagnemma and Dubowsky, 2004] Iagnemma, K. and Dubowsky, S. (2004). *Mobile Robots in Rough Terrain: Estimation, Motion Planning, and Control with Application to Planetary Rovers.* Springer Verlag.

[Iglesias et al., 1998] Iglesias, R., Regueiro, C. V., Correa, J., Schez, E., and Barro, S. (1998). Improving wall following behaviour in a mobile robot using reinforcement learning. In *Proc. of the International ICSC Symposium on Engineering of Intelligent Systems.* ICSC Academic Press.

[Kantz and Schreiber, 1997] Kantz, H. and Schreiber, T. (1997). *Nonlinear time series analysis.* Cambridge University Press, Cambridge.

[Kantz and Schreiber, 2003] Kantz, H. and Schreiber, T. (2003). *Tisean — Nonlinear time series analysis.* http://www.mpipkd-dresden.mpg.de/tisean.

[Kaplan and Glass, 1995] Kaplan, D. and Glass, D. (1995). *Understanding nonlinear dynamics.* Springer Verlag, New York.

[Katevas, 2001] Katevas, N., (Ed.) (2001). *Mobile Robotics in Health Care Services.* IOS Press.

[Kennel and Isabelle, 1992] Kennel, M. and Isabelle, S. (1992). Method to distinguish possible chaos from colored noise and to determine embedding parameters. *Phys. Rev. A*, 46:3111–3118.

[Kennel et al., 1992] Kennel, M. B., Brown, R., and Abarbanel, H. D. I. (1992). Determining embedding dimension for phase-space reconstruction using a geometrical construction. *Physical Review A*, 45:3403–3411.

[Klayman and Ha, 1987] Klayman, J. and Ha, Y. (1987). Confirmation, disconfirmation and information in hypothesis testing. *Psychological Review*, 94:211–228.

[Korenberg et al., 1988] Korenberg, M., Billings, S., Liu, Y., and McIlroy, P. (1988). Orthogonal paramter estimation for non-linear stochastic systems. *Int. J. Control*, 48:193–210.

[Kube and Zhang, 1992] Kube, C. and Zhang, H. (1992). Collective robotic intelligence. In Meyer, J., Roitblat, H., and Wilson, S., (Eds.), *From Animals to Animats 2*, Cambridge MA. MIT Press.

[Kuhn, 1964] Kuhn, T. (1964). *The structure of scientific revolutions.* University of Chicago Press, Chicago.

[Kurz, 1994] Kurz, A. (1994). *Lernende Steuerung eines autonomen mobilen Roboters.* VDI Verlag, Düsseldorf.

[Levin and Rubin, 1980] Levin, R. and Rubin, D. (1980). *Applied elementary statistics.* Prentice-Hall, Englewood Cliffs.

[Ljung, 1987] Ljung, L. (1987). *System identification : theory for the user.* Prentice-Hall.

[Loeb, 1918] Loeb, J. (1918). *Forced movements, tropisms and animal conduct.* J.B. Lippencott, Philadelphia.

[Mañe, 1981] Mañe, R. (1981). On the dimension of the compact invariant set of certain nonlinear maps. In *Lecture Notes in Mathematics 898*, pages 230–242, Berlin, Heidelberg, New York. Springer Verlag.

[Martin, 2001] Martin, F. G. (2001). *Robotic explorations, a hands-on introduction to engineering.* Prentice Hall.

[Mataric, 1994] Mataric, M. (1994). Learning to behave socially. In Cliff, D., Husbands, P., Meyer, J., and Wilson, S., (Eds.), *From Animals to Animats 3*, Cambridge MA. MIT Press.

[McKerrow, 1991] McKerrow, P. (1991). *Introduction to Robotics.* Addison-Wesley, Sydney.

[Merriam Webster, 2005] Merriam Webster (2005). Online dictionary. http://www.m-w.com/.

[Morik, 1999] Morik, K., (Ed.) (1999). *Making Robots Smarter: Combining Sensing and Action Through Robot Learning.* Kluwer Academic Publishers.

[Murphy, 2000] Murphy, R. (2000). *Introduction to AI Robotics.* MIT Press, Cambridge, Mass.

[Nehmzow, 2003a] Nehmzow, U. (2003a). *Mobile Robotics: A Practical Introduction.* Springer, Berlin, Heidelberg, London, New York.

[Nehmzow, 2003b] Nehmzow, U. (2003b). Navigation. *Encyclopedia of Cognitive Science.*

[Nola and Sankey, 2000] Nola, R. and Sankey, H. (2000). A selective survey of theories of scientific method. In Nola, R. and Sankey, H., (Eds.), *After Popper, Kuhn and Feyerabend*, Dordrecht. Kluwer.

[Oreskes et al., 1994] Oreskes, N., Shrader-Frechette, K., and Belitz, K. (1994). Verification, validation and confirmation of numerical models in the earth sciences. *Science*, 263.

[Parker, 1994] Parker, L. (1994). *Heterogeneous multi-robot cooperation.* PhD thesis, Massachussetts Institute of Technology, Department of Electrical Engineering and Computer Science.

[Paul and Elder, 2004] Paul, R. and Elder, L. (2004). *The miniature guide to critical thinking.* Foundation for Critical Thinking, Dillon Beach CA.

[Pearson, 1999] Pearson, R. (1999). *Discrete-time dynamic models.* Oxford University Press, Oxford.

[Peitgen et al., 1992] Peitgen, H., Jürgens, H., and Saupe, D. (1992). *Chaos and fractals — new frontiers of science.* Springer Verlag, New York, Berlin, Heidelberg, London.

[Pena et al., 2001] Pena, D., Tiao, G., and Tsay, R., (Eds.) (2001). *A course in time series analysis.* Wiley, New York.

[Popper, 1959] Popper, K. (1959). *The logic of scientific discovery.* Hutchinson, London.

[Popper, 1963] Popper, K. (1963). *Conjectures and refutations: the growth of scientific knowledge.* Routledge and K. Paul, London.

[Popper, 1972] Popper, K. (1972). *Objective knowledge.* Clarendon Press, Oxford.

[Ritter et al., 2000] Ritter, H., Cruse, H., and Dean, J., (Eds.) (2000). *Prerational intelligence*, Dordrecht. Kluwer.

[Rosenstein et al., 1994] Rosenstein, M. T., Collins, J. J., and De Luca, C. J. (1994). Reconstruction expansion as a geometry-based framework for choosing proper delay times. *Physica D*, 73:82–98.

[Sachs, 1982] Sachs, L. (1982). *Applied statistics.* Springer Verlag, Berlin, Heidelberg, New York.

[Schöner and Kelso, 1988] Schöner, G. and Kelso, J. (1988). Dynamic pattern generation in behavioral and neural systems. *Science*, 239:1513–1520.

[Scilab Consortium, 2004] Scilab Consortium (1989–2004). The Scilab programming language. *http://www.scilab.org*.

[Siegwart and Nourbakhsh, 2004] Siegwart, R. and Nourbakhsh, I. R. (2004). *Introduction to Autonomous Mobile Robots*. MIT Press.

[Smithers, 1995] Smithers, T. (1995). On quantitative performance measures of robot behaviour. *Robotics and Autonomous Systems*, 15(1-2):107–133.

[Sobol, 1993] Sobol, I. (1993). Sensitivity estimates for nonlinear mathematical models. *Mathematical Modelling and Computational Experiment*, 1:407–414.

[Steels, 1995] Steels, L., (Ed.) (1995). *The biology and technology of intelligent autonomous agents*, Berlin, Heidelberg, New York. Springer Verlag.

[Takens, 1981] Takens, F. (1981). Detecting strange attractors in turbulence. In *Dynamical Systems and Turbulence*, pages 366–381, Berlin, Heidelberg, New York. Springer Verlag.

[Theiler and Lookman, 1993] Theiler, J. and Lookman, T. (1993). Statistical error in a chord estimator of the correlation dimension: the rule of 'five'. *Bifurcation and Chaos*, 3:765–771.

[Ueyama et al., 1992] Ueyama, T., Fukuda, T., and Arai, F. (1992). Configuration and communication structure for distributed intelligent robot system. *IEEE Trans. Robotics and Automation*, pages 807–812.

[von Randow, 1997] von Randow, G. (1997). *Roboter*. Rowohlt, Reinbek.

[Walter, 1950] Walter, W. G. (1950). An imitation of life. *Scientific American*, 182:43–45.

[Walter, 1951] Walter, W. G. (1951). A machine that learns. *Scientific American*, 51:60–63.

[Wilcoxon, 1947] Wilcoxon, F. (1947). Probability tables for individual comparisons by ranking methods. *Biometrics*, 3:119–122.

[Wiltschko and Wiltschko, 2003] Wiltschko, R. and Wiltschko, W. (2003). Avian navigation: from histgorical to modern concepts. *Animal Behaviour*, 65:257–272.

[Wolf, 2003] Wolf, A. (2003). Chaos analysis software. http://www.cooper.edu/~wolf/chaos/chaos.htm.

[Wolf et al., 1995] Wolf, A., Swift, J., Swinney, H., and Vastano, J. (1995). Determining lyapunov exponents from a time series. *Physica 16D*.

[Wyatt and Demiris, 2000] Wyatt, J. and Demiris, J., (Eds.) (2000). *Advances in Robot Learning: 8th European Workhop on Learning Robots*. Springer-Verlag.

Index

χ^2 analysis, program, 73
χ^2 test, 70
χ^2 test, example, 72

Analytical robotics, 13, 14, 199, 200
ANOVA, non-parametric, 53
ANOVA, parametric, 42
ANOVA, parametric example, 43
ANOVA, parametric, testing for significance, 43
Aperiodicity, 116
ARMAX, 150, 151
ARMAX, Scilab code, 152
Association between nominal variables, 73
Association between nominal variables (entropy), 75
Attractor reconstruction, 90
Attractor reconstruction, example, 93
Attractor, definition, 86
Attractor, dimension, 116
Autocorrelation, 142

Bacon, F., 20
Baselines in experimental design, 25
Behaviour dynamics, comparison, 190
Behaviour, emergence, 5
Blind experimentation, 27
Boundedness, 98

Carrier pigeon, 133
Categorical data, χ^2 test, 70
Categorical data, analysis, 69
Ceiling effect, 26
Chaos walker, 126

Chaos walker, attractor, 130
Chaos, deterministic, 197
Confidence interval for mean, 32
Confirmation bias, 22
Confounding effects, 25
Conspiracy of goodwill, 26
Constant errors, 26, 27
Contingency table analysis, 69
Contingency tables, 69
Controls in experimental design, 26
Cooperative robots, 4
Correlation analysis, parametric, 57
Correlation coefficient, linear, 62
Correlation coefficient, Pearson, 62
Correlation coefficient, testing for significance, 63
Correlation dimension, 116
Correlation distance, 116
Correlation integral, 116
Counterbalancing, 27
Cramer's V, 73
Cramer's V, program, 73
Cramer's V: example, 74
Crosstabulation analysis, 69

Degrees of freedom (mobile robot), 87
Degrees of freedom, χ^2 analysis, 71
Description, quantitative, 200
Deterministic chaos, 197
Deterministic signal, 97
Dimension of attractors, 116
Dimension, correlation, 116
Door traversal, 181
Down sampling, 141

Dynamical systems, 85
Dynamical systems theory, 11, 85, 197

Embedding dimension, 92
Embedding lag, 92
Embedding lag, determination, 93
Entertainment robots, 3
Entropy, 75, 76
Errors, constant, 27
Existence proof, x, 14, 17

F-statistic, 61
F-statistic, critical values table, 62
False nearest neighbours, 92, 103
Falsificationism, 20
Floor effect, 26
Frequency spectrum, 142

Gaussian distribution, 30, 31
Gaussian distribution, table, 33

Hardwired control program, 146
Health care robots, 3
Hypothesis, 21
Hypothesis, causal, 22
Hypothesis, descriptive, 22

Induction, 20
Induction, problem, 20
Inductivism, 20
Information loss, estimation, 105
Information, mutual, 93
Inspection robots, 3
Iterative refinement, 14, 19

Kruskal Wallis test, 53
Kruskal Wallis test, example, 54
Kuhn, T., 20

Learning controller, simulation, 148
Linear correlation coefficient, 62
Linear correlation, Scilab code, 63
Linear regression, 57, 58
Linear regression Scilab, 59
Linear regression, testing for significance, 60
Logistic map, 108
Lyapunov exponent, 100
Lyapunov exponent, estimate from a time series, 101

Lyapunov exponent, estimation from information loss, 113
Lyapunov exponent, robotics example, 102

Magellan Pro robot, 2
Mann-Whitney U-test, 45
Mean, 30, 31
Model, meaning, 200
Mopping up, 21
Multilayer Perceptron, 143
Museum tour guide robot, 3
Mutual information, 93, 142

NARMAX, 150, 155
Nominal variables, 69
Nominal variables, association, 70
Non-parametric ANOVA, 53
Non-parametric methods, 43
Non-parametric tests for a trend, 65
Non-stationary data, making stationary, 100
Normal distribution, 30, 31, 36
Normal distribution, table, 33
Normal probability paper, 36
Normal science, 20
Null hypothesis, 29

Obstacle avoidance, phase space reconstruction, 94
Occam's razor, 23
Orbit, 86, 100

Paired samples, parametric example, 40
Paradigm, scientific, 20
PCA, 80
Pearson correlation coefficient, 62
Pearson's r, 62
Pearson's r, Scilab code, 63
Pearson's r, significance, 63
Perceptron, multilayer, 143
Phase space, 85, 86
Phase space reconstruction, 90
Phase space reconstruction, example, 93
Pigeon, 133
Platform-independent programming, 179
Popper, K., 20
Prediction horizon, 106, 190
Prediction horizon, example, 111
Prediction of robot behaviour, 146
Prediction of sensory perception, 146

Principal component analysis, 80
Problem of induction, 20
Pseudoreplication, 26

Quadratic iterator, 108
Quantitative descriptions, role, 17

Randomisation, 27
Randomness, 55
Rank correlation, non-parametric, 65
Rank correlation, Spearman, 65
Rank correlation, testing for significance, 66
Regression, linear, 57, 58
Regression, linear (testing for significance, 60
Repeatability of robot behaviour, 6
Replication of experiments, 17
Return plots, 98
Revolution, scientific, 20
Robot behaviour, prediction, 146
Robot engineering, 18
Robot identification, 156, 193, 198
Robot Java, 179
Robot science, 18
Robot training, 181
Robot-environment interaction, dynamic analysis, 197
Robot-environment interaction, static analysis, 196
Robot-environment interaction, theory, 195, 197
RobotMODIC, 179
Runs test, 55, 99

Scaling region, 102, 117
Scientific paradigm, 20
Scientific research methodology, 21
Scientific revolution, 20
Scilab, 27
Self-localisation through environment identification, 162
Sensitivity analysis, 200
Sensitivity to initial conditions, 100
Sensor identification, 184
Sensor identification, example, 185

Sensory perception, prediction, 146
Significance level, 33, 34
Simulation, advantages, 140
Simulation, limitations, 200
Slaving principle, 140
Spearman rank correlation, 65
Spearman rank correlation, example, 65
Spearman rank correlation, Scilab code, 68
Spearman rank correlation, testing for significance, 66
Standard deviation, 30, 32
Standard error, 30, 32
State space, 85
Stationarity, 98
Stochastic signal, 97
Surveillance robots, 3
System identification, 11, 140, 193, 198

T-test, 61
T-test for independent samples, example, 39
T-test for paired data, example, 40
T-test, dependent samples, 40
T-test, dependent samples example, 40
T-test, in linear regression, 61
T-test, independent samples, 38
Theory of robot-environment interaction, 16, 197
Theory, definition, 16
Time lag embedding, 92
Training of robots, 181
Trend, non-parametric tests, 65
Trend, testing for, 57
Trial and error procedures, 19
Type I error, 34
Type II error, 34

U-test, 45
Uncertainty coefficient, 76, 78
Uncertainty coefficient, example, 78
Uncertainty coefficient, program, 79

Verification, independent, 17

Wilcoxon rank sum test, 45
Wilcoxon test, paired observations, 50